AF443061

Optogalvanic Effect in Ionized Gas

N.G. Preobrazhensky
1933-1992

Optogalvanic Effect in Ionized Gas

V.N. Ochkin, *P.N. Lebedev Physical Institute, Moscow*
N.G. Preobrazhensky, *Institute for Theoretical and Applied Mechanics, Novosibirsk*
N.Y. Shaparev, *Krasnoyarsk Scientific Centre, Krasnoyarsk*

Translated from the Russian by Sergei G. Kittel

LEBEDEV PHYSICAL INSTITUTE UNIVERSITY PRESS
FOUNDATION FOR INTERNATIONAL SCIENTIFIC AND EDUCATION COOPERATION
London • Moscow

Originally published in Russian in 1991as
ОПТОГАЛЬВАНИЧЕСКИЙ ЭФФЕКТ В ИОНИЗОВАННОМ ГАЗЕ
by Nauka, Moscow
© 1991 Nauka, Moscow

Amsteldijk 166
1st Floor
1079 LH Amsterdam
The Netherlands

British Library Cataloguing in Publication Data

Ochkin, V. N. (Vladimir Nikolaevich)
   Optogalvanic effect in ionized gas
    1.Optogalvanic spectroscopy 2.Ionized gases
    I.Title II.Preobrazhensky, N. G. III.Shaparev, N. Y.
   530.4′46

ISBN: 90-6994-001-9

Contents

Preface

Though the optogalvanic effect has been known about for over sixty years, the development of the underlying theory and its practical applications has been extremely sporadic. This is evidenced by the fact that there have been almost as many papers published in scientific journals over the past decade than were published during the fifty years immediately following the discovery of the effect. This phenomenon is largely due to recent progress made in quantum electronics (specifically tunable laser physics and technology) and gas-discharge physics.

At the same time, despite the rapid increase in the number of papers published, until now no monograph has been written on the subject. One reason for this absence is that writing such a monograph is not a very rewarding task, for the material dates extremely quickly. This fact notwithstanding, the authors who ventured to publish a review on the optogalvanic effect in *Uspekhi Fizicheskikh Nauk* (Advances in Physical Sciences) in 1966 (Vol. 148, Issue 3), have now also dared to publish this book.

Without attempting to perform the impossible, we have used the ocean of literature dealing with the optogalvanic effect in gases and plasma and selected the principal questions which have already been settled to some degree. We have also attempted to characterize the current theoretical models which provide an acceptable quantitative agreement between theoretical and experimental data. In many cases these models are rather complex and require rigorous calculations. A substantial amount of these calculations and theoretical model investigations have been performed by the authors with their co-workers.

The applications of the optogalvanic effect are exceptionally interesting and diverse. There are new, highly sensitive express spectrochemical analysis techniques (both atomic and molecular), high-precision laser frequency stabilization, plasma diagnostics, studies of the burring process mechanism, and isotope separation. The authors have tried to illustrate these and other possibilities by including some examples and listing the relevant references. If the book should prove helpful in recruiting young, vigorous specialists for research in this promising field, the authors will consider their objective met.

The authors would like to take this opportunity to express their deep gratitude to N.K. Zaitsev, V.A. Pushkarev, Yu.B. Udalov and S.N. Tskhai for the help they rendered at various stages of the work. Special thanks also to A.E. Bulyshev and N.V. Denisova for their kind permission to include their fresh (and frequently as yet unpublished) theoretical materials in Chapters 2 and 3. We are also indebted to I.M. Beterov, V.S. Vorobyov, N.B. Zorov, Yu.Ya. Kuzyakov, B.M. Smirnov and E.A. Yukov for many useful discussions.

Nomenclature

I – light intensity

A, A' – spontaneous decay probability and spontaneous decay probability with due regard for radiation trapping

σ_{ph} – photoexcitation cross section

F – photon flux

γ – de-excitation probability

k_0 – line-center absorption coefficient

c – velocity of light

B_{12} – Einstein's absorption coefficient

Q – number of photons absorbed, generation power

ω_{21} – transition frequency

W – spontaneous loss of atoms with due regard for radiation trapping and diffusion

$\mathbf{k}$ – wave vector

Γ_0 – homogeneous half-width

Ω – frequency detuning

P^{ph} – radiant excitation rate

i, j – electron current, electron current density

i_i, j_i – ion current, ion current density

z – ballast resistance

E – electric field strength

ε – supply source e.m.f., electron energy

R_e, R_s, R_d – external circuit resistance, supply source resistance, and discharge resistance, respectively

α_c – cathode region resistance

V_a, V_c – anode and cathode voltage drops, respectively

σ – conductivity

f – volt-ampere characteristic

n_a – total atomic density

n_k – atomic state population

S^{at} – atomic excitation and quenching rate constants

T_e, T_g – electron and gas temperatures, respectively

$W_{se, el, ex, i}$ – energy losses

a – associative ionization rate

P – total electronic excitation rate, electric power

Q' – energy deposition into high-frequency discharge

ε_e – energy lost by electron upon collision

τ_m – time metastable atoms take to reach the wall by diffusion

τ_α – ambipolar diffusion time

v_t – transport electron collision frequency

D^+ – ambipolar diffusion coefficient

μ_e, μ_i – electron and ion mobilities, respectively

λ_e – electron free path length

D_m – metastable atom diffusion coefficient

f_0 – electron velocity distribution function

$\tau_{||}$ – ion drift time in cathode region

l – cathode region length

v_e – electron drift velocity

k_T – thermal conductivity coefficient

g_k – statistical weight of the kth state

l, R – cylindrical tube discharge length and radius, respectively

k – Boltzmann constant

H – magnetic field strength

e, m_e – electronic charge and mass, respectively

G – right-hand side of electron-balance equation

K – right-hand side of atom-balance equation

E_{mk} – energy interval between levels m and k

I_i – ionization potential of the ith state

n_e – electron concentration

n_i – ion concentration

S – electronic excitation, quenching, and ionization rate constants

p – gas pressure

M_i – ionic mass

$v_d^{(i)}$ – ion drift velocity

v – atomic velocity

F_p – light pressure force

c_s – ion-sound wave velocity

Chapter 1

Background of the Optogalvanic Effect

It is well known that absorption spectroscopy is a powerful tool for studying atomic and molecular structures and elementary processes, determining concentrations of substances, diagnosing various objects, and so on. Quantitative measurements of the density of particles in various quantum states are in that case based on the determination of the quantity ΔI – the change in the intensity of light which has passed through the object of interest. If the entrant radiation has a high intensity, its influence on the object must be taken into account, but the quantity being measured is still the light intensity variation ΔI.

An alternative may be the measurement of the variation of some parameter of the object, caused by the light passing through it. As applied to gaseous and plasma objects, laser-induced fluorescence and acousto-optic spectroscopy techniques have been widely recognized. The light absorbed by a gas causes either its excess fluorescence or some changes in its density. A description of the essence of these techniques and their main applications can be found, for example, in [1.1] and references cited therein. The same work also touches upon another method based on light-induced changes in the electrical characteristics of the object. In various works, this effect has been called the optogalvanic effect, optovoltaic effect, optoelectric effect, or the ionization enhancement effect. We will adhere to the first name. The principle of the effect is that the absorption of light causes a redistribution of level populations in atoms or molecules. As a result of various radiative and (or) collisional, and (or) collective processes, changes take place in the density of charged particles, their mobility, and energy, which affects the ionization balance and conductivity of the object. Ionization occurs as a consequence of additional (with respect to absorption) processes. The relationship between the magnitude of the effect and the radiation wavelength is, in that case, of resonant character. Conductivity may also vary simply as a result of photoprocesses, such as, photoionization (multiphoton and stepwise types included) [1.2, 1.3], or because of nonresonant absorption [1.4]. In the final analysis, the optogalvanic effect is detected by monitoring the change of the current, Δi, in or voltage, ΔV, across the electric circuit including the object under study. The ratio between the change in the power dissipated in an external circuit and the absorbed light power is also used as a measure of the optogalvanic effect. Such observations are most natural for gas-discharge objects, and it is here that they have gained acceptance. The optogalvanic effect was observed for the first time by Foote and Mohler [1.5] in gases and Penning [1.6] in a gas-discharge plasma. When illuminating a 20-Torr neon discharge with the emission from a similar discharge, the author of [1.6] observed an increase of some 10 % in the voltage across the discharge tube. To demonstrate the effect graphically, the discharge tube was connected in a circuit containing a capacitor and a loudspeaker – the pitch of the sound changed the instant the discharge was illuminated (Fig. 1.1). In later works, the effect was discussed and studied mainly with a view to revealing the role of metastable states in the stepwise ionization processes [1.7–1.9].

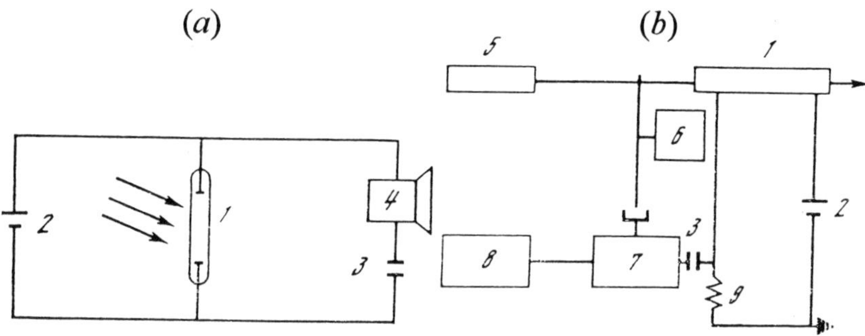

Figure 1.1. Detection of the optogalvanic effect. (a) Schematic diagram of the Penning experiment on the change of the sound pitch of a loudspeaker upon illumination of a discharge; (b) block diagram of a measuring system using a laser. 1 – discharge; 2 – high-voltage source; 3 – capacitor; 4 – loudspeaker; 5 – laser; 6 – modulator; 7 – phase-sensitive detector; 8 – recorder; 9 – resistor.

The optogalvanic effect discovered in [1.6] was used to verify the presence of what is known as the Penning ionization mechanism:

$$Ne^m + Ar \rightarrow Ne + Ar^+ + e, \qquad (1.1)$$

where Ne^m is the group of metastable neon states with an energy higher than the ionization energy of the argon atom. When illuminating the discharge, the neon atoms in the Ne^m states rise to highly excited states which may decay by channels passing the Ne^m states, thus slowing reaction (1.1). However, other experiments showed that the optogalvanic effect occurred in pure neon discharges as well, which apparently should be associated with the weakening of the mechanism of stepwise electronic ionization via the Ne^m states under irradiation. In the next works [1.7, 1.8], experiments were conducted with a spectral resolution and showed the optogalvanic effect to be dependent on the discharge conditions and the type of the optical transition involved. In all the experiments [1.6–1.8], the effect was observed to be positive (i. e., it corresponded to the rise of the discharge voltage, $\Delta V > 0$, with the discharge current remaining fixed).

Recent investigations using lasers have uncovered a number of new features of the phenomenon and greatly increased the range objects studied. In describing the mechanism of the optogalvanic effect, one can roughly single out two general stages. First, it is necessary to find out how the absorption of light can cause changes in the density and (or) energy of charged particles. Secondly, one has to associate these changes with variations of the macroscopic characteristics of the object and the parameters of the electric circuit. The first part of the task requires using a kinetic approach, and the second presupposes the knowledge of the model of the object and discharge circuit reaction to perturbations in the density and mobility of charges, though, of course, both these parts are interrelated.

Due to the needs of absorption spectroscopy, the detection of optogalvanic signals in gases and plasmas has rapidly been perfected, and it soon became clear that the optogalvanic effect can be used not only as a convenient means of detecting absorption, but also as a fairly universal method for studying the characteristics of gaseous and plasma media.

The interest in optogalvanic phenomena has advanced not only due to the advent of lasers, but also in connection with the developmental needs of laser technology itself. The optogalvanic effect has been used to study the processes taking place in the active media of gas-discharge lasers and to stabilize laser frequencies. Another avenue of research was created by the authors of [1.10, 1.11] and then explored more actively in [1.12–1.14]. This is connected with the use of tunable lasers for spectroscopic purposes. One of the main features of laser optogalvanic spectroscopy is the simplicity of detecting the signal: the latter is simply picked by a phase-sensitive detector off a resistor in the discharge supply circuit (Fig. 1.1b). A capacitor is used in the detector circuit to block the high d. c. discharge supply voltage.[1] Use in that case is made of continuous-wave or pulsed dye lasers [1.12, 1.16], color-center lasers [1.17, 1.18], and diode lasers [1.19], depending on the spectral region to be covered and experimental requirements. Frequency-doubled laser schemes [1.20] are also in use.

Figure 1.2a shows an example of the optogalvanic spectrum of a cesium discharge irradiated with a dye laser in the region 800–890 nm [1.21]. Note the high signal/noise ratio even for very weak transitions.

Figure 1.2b presents the spectra of a He-Ne discharge recorded upon irradiation with a dye laser in the region 565–645 nm [1.22]. The spectrum A represents the ordinary absorption of the laser power, and the spectrum B was recorded by means of the optogalvanic detection technique. As one can see, the optogalvanic method is highly sensitive. At the same time, it follows from Fig. 1.2b that the optogalvanic spectrum is clearly distinct from its ordinary optical counterpart. Whereas all the lines in the optical spectrum correspond to the absorption of light, the quantity ΔV in the optogalvanic spectrum differs in sign between different lines. Obviously to interpret such a spectrum requires a detailed analysis of its formation.

Many investigations have recently been conducted into both the nature of the optogalvanic effect and its applications and some general works on the individual aspects of the problem have also been done (see, for example, [1.23–1.28]). The purpose of this book is to give the reader an idea of the current status of research into the laser-induced optogalvanic phenomena in gases and plasmas and their potential applications.

---

[1] In more recent works using lasers, some authors used a method which was similar to the one used by Penning for demonstration purposes (Fig. 1a) but involved more accurate measurements of the oscillation frequency of the relaxation generator [1.15]. Oddly enough, this led to a substantial improvement of sensitivity in comparison with synchronous detection.

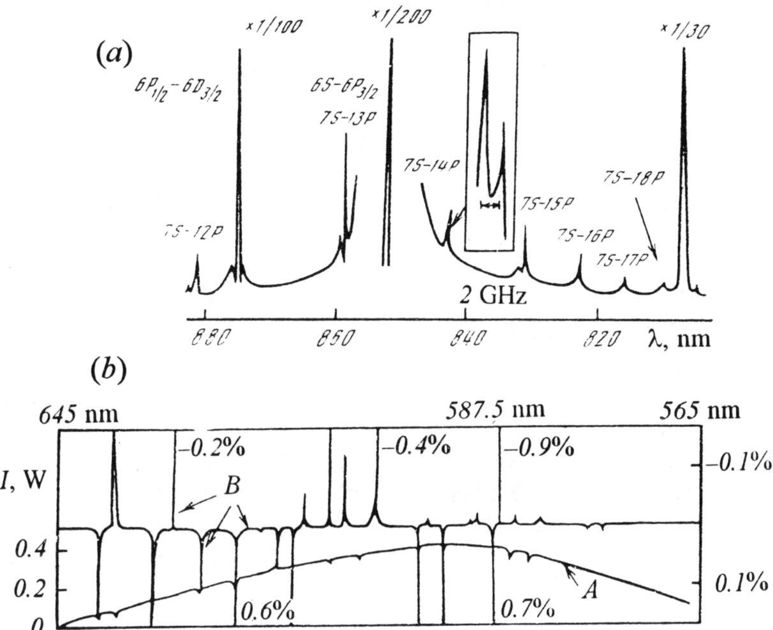

Figure 1.2. Exemplary optogalvanic spectra. (a) Optogalvanic spectrum of cesium [1.21], voltage variations; (b) spectra of a He-Ne discharge [1.22]. Y-axis: left – radiation power passed through the discharge; right – relative voltage variation. A – optical spectrum; B – optogalvanic spectrum. For spectral peaks extending beyond the figure field, the magnitude of the optogalvanic effect is indicated as a relative voltage change (%).

REFERENCES

1.1. N. Omenetto, ed., *Analytical Laser Spectroscopy* (New York: Wiley, (1979).

1.2. B. H. Rockney, T. A. Cool, E. R. Grant, *J. Chem Phys.* **87**: 141–144 (1982).

1.3. W. G. Mallard, J. H. Miller, K. G. Smith, *J. Chem. Phys.* **76**: 3483 – 3490 (1983).

1.4. N.K. Zaitsev, N.Ya. Shaparev, *Optoelektricheskie yavleniyav plazme* (Photoelectric phenomena in plasma) (Krasnoyarsk: L. V. Kirenskiy Physical Institute, Preprint No. 207–209F, 1982) (in Russian).

1.5. P. D. Foote, F. L. Mohler, *Phys. Rev.* **26**: 195–207 (1925).

1.6. F. M. Penning, *Physica* **8**: 137–140 (1928).

1.7. C. Kenty, *Phys. Rev.* **80**: 95–98 (1950).

1.8. K. W. Meissner, W. F. Miller, *Phys. Rev.* **92**: 868–898 (1953).

1.9. Yu. Z. Ionikh, N. P. Penkin, A. V. Samson, *Vestnik LGU. Fizika. Khimiya.* No. 4, iss. 1: 51–60 (1976).

1.10. C. B. Collins, B. W. Johnson, D. Popescu *et al.*, *Phys. Rev.* **8**: 2179–2201 (1973).

1.11. D. Popescu, M.L. Pascu, C.B. Collins *et al.*, *Phys. Rev.* **8**: 1666–1672 (1973).

1.12. R. B. Green, R. A. Keller, G. G. Luther *et al.*, *Appl. Phys. Lett.* **29**: 727–731 (1976).

1.13. W. B. Bridges, *JOSA* **68**: 352–360 (1978).

1.14. N.K. Zaitsev, N.Ya. Shaparev, Author's Certificate No. 678948, *Bull. Izobreteniy* No. 29: 283 (1981).

1.15. G.-Y. Yan, K.-I. Fujii, A. L. Schawlow, *Opt. Lett.* **52**, No. 2: 142–144 (1990).

1.16. R. Vasudev, R. N. Zare, *J. Chem. Phys.* **76**: 5276–5270 (1982).

1.17. S. A. M. Al-Chalabi, R. S. Stewart, R. Illingworth, I. S. Ruddock, *J. Phys. D.* **16**: 115–124 (1980).

1.18. D. J. Jackson, E. Arimondo, J. E. Lawler, T.W. Hänsch, *Opt. Comm.* **33**: 51–55 (1980).

1.19. R. C. Webster, R. T. Mensies, *J. Chem. Phys.* **78**: 2121–2128 (1983).

1.20. J. E. M. Goldsmith, *J. Chem. Phys.* **78**: 1610–1611 (1983).

1.21. L. Ph. Rosch, *Opt. Comm.* **44**: 259–261 (1983).

1.22. T. F. Johnston, *Laser Focus* **14**: 58 (1978).

1.23. R. C. Webster, C. T. Rettner, *Laser Focus* **19**: 41–52 (1983).

1.24. J. E. M Goldsmith, J.E. Lawler, *Contemp. Phys.* **22**: 235–248 (1981).

1.25. A. I. Ferguson, *Phil. Trans. R. Soc. Lond. A.* **307**: 645–657 (1982).

1.26. V. N. Ochkin, N. G. Preobrazhensky, N.N. Sobolev, N.Ya. Shaparev, *UFN* **148**, No. 3: 473–507 (1986).

1.27. B. Barberi, N. Beverini, A. Sasso, *Rev. Mod. Phys.* **62**, No. 3: 603–645 (1990).

1.28. R. S. Stewart, J. E. Lawler, ed., *Proceedings of Conference on Optogalvanic Spectroscopy, IOP Conference Series* **113** (1991)

Chapter 2

Optogalvanic Effect in Glow Discharges

2.1 GENERAL REMARKS

In the available literature of the optogalvanic effect in glow discharges differing in geometry discharge tubes 2–10 mm in diameter and around 100 mm in positive column length are used. The discharge current ranges between 1 and 100 mA, and gas pressure, from 0.1 to 5 Torr. The resistance z of the ballast resistor, which also doubles as an instrument resistor, is 1–100 kΩ, and the d. c. source voltage is between 100 and1000 V. To make the light beam pass through the positive column only, the electrodes are usually arranged in offsets, but coaxial discharge tube designs are also not uncommon. The magnitude of the a. c. voltage resulting from a modulated radiation around 10^{-1} W in power being passed through the discharge typically amounts to some 10^{-2} V. Longer discharges (for example, 300 mm [1.16]) are in use as well. The optogalvanic effect is also observed when irradiating the discharge in a transverse direction (see, for example, [2.1]).

A glow discharge consists of electrode regions and a positive column. And it is usually in the positive column region that the gas is excited by the radiation. In that case, changes occur in the conductivity of this region and in the electric field strength, ΔE, in it. At the same time, the response of the discharge current, whose magnitude depends on the type of the discharge and its electric circuit parameters, occurs.

The magnitude of the current flowing through the electric circuit including the gas-discharge device is defined by Ohm's law for a closed circuit:

$$i = \varepsilon / (R_e + R_s + R_d),\tag{2.1}$$

where ε is the e. m. f. of the supply source and R_e, R_s, and R_d are the resistances of the external circuit, supply source, and discharge, respectively.

Consider the relationship between the change in the current flowing in the glow discharge circuit and that in the electric field strength in the positive column of the discharge.

Let us single out two types of discharge: (1) normal glow discharge wherein the cathode glow voltage and current density remain constant irrespective of the discharge current variations and (2) abnormal glow discharge wherein the cathode glow voltage grows higher with increasing current density [2.2].

1. Normal Glow Discharge. The voltage drop across the discharge gap, V, is equal to the sum of the voltage drops across the individual sections of the discharge:

$$V = V_a + V_c + EL,\tag{2.2}$$

where A_a and V_c are the anode and cathode voltage drops, respectively, and L is the length of the positive column.

In a normal glow discharge, V_a and V_c remain unchanged no matter what the discharge current variations. In that case, expression (2.1) may be written as

$$i = (V_1 - EL) / R_1,\tag{2.3}$$

where $V_1 = \varepsilon - V_a - V_c$, $R_1 = R_e + R_s$, and $V_1 - EL \equiv \varepsilon - V = \varepsilon - iR_d$. The change of the discharge current, Δi, as a function of the light-induced electric field variation ΔE is defined by the expression

$$\Delta i = -(L / R_1)\Delta E.\tag{2.4.}$$

If the supply source resistance $R_s \ll R_e$, then $R_1 \approx R_e \approx z$, and so

$$\Delta i = -(L / z)\Delta E. \tag{2.5}$$

2. Abnormal Glow Discharge. In that case, the anode voltage drop $V_a =$ const, whereas the voltage drop in the cathode region depends on the discharge current: $V_c = V_c(i)$. If the change $V_c(i)$ under external irradiation is small, it then may be expressed to a first approximation in the form

$$V_c(i) \cong V_c(i_0) + \alpha_c(i - i_0), \quad \alpha_c = \left.\frac{\partial V_c}{\partial i}\right|_{i=i_0}, \tag{2.6}$$

where i_0 is the steady-state current through the discharge prior to its being exposed to the external radiation. Considering (2.6) and subject to the condition that the anode voltage drop remains unchanged, Ohm's law for the closed circuit will assume the following form:

$$i = (V_2 - EL) / R_2, \tag{2.7}$$

where

$$V_2 = \varepsilon - V_a - V_c(i_0) - \alpha_c i_0 \text{ and } R_2 = R_e + R_i + \alpha_c. \tag{2.8}$$

In that case,

$$\Delta i = -(L / R_2)\Delta E. \tag{2.9}$$

If the voltage across the discharge tube remains the same, which holds true subject to the condition $\alpha_c \gg R_e, R_s$, the current response is

$$\Delta i = -(L / \alpha_c)\Delta E. \tag{2.10}$$

Comparing the optogalvanic current responses of the abnormal and normal glow discharges, (2.9) and (2.4), one can see that under small perturbation conditions the relationship between Δi and ΔE is in both cases of linear character. Therefore, one can perform numerical calculations for the electric field strength variation ΔE in the positive column and then find Δi either from expression (2.4) or from expression (2.9), depending on the type of discharge in hand.

The light-induced changes in the positive column characteristics of the discharge may be described by means of the following set of equations.

The populations of the excited states are defined by the following expression:

$$\frac{dn_k}{dt} \equiv K_k = n_e \left\{ \sum_{m \neq k} \left[S_{mk} n_m - (S_{km} + S_k) n_k \right] \right\}$$

$$+ n_1 \left[\sum_{m \neq k} \left(S'^{at}_{mk} n_m - S'^{at}_{km} n_k \right) \right] \qquad (2.11)$$

$$+ \sum_{m > k} A'_{mk} n_m - n_k \sum_{1 \leq l \leq k} A'_{kl} - \tau_m^{-1} n_k - F \sigma_{ph} \left(n_k - \frac{g_m}{g_k} n_m \right),$$

where n_k and n_m are the populations of the kth and mth levels, respectively, such that $1 \leq m \leq r$, $1 \leq l < k$, and $2 \leq k \leq r$, r is the number of energy levels considered, n_e is the electron concentration, S_{mk} is the rate at which atoms move from the mth to the kth state, S_{km} is the rate at which atoms drop from the kth state back to the mth state as a result of electron-atom collisions, S'^{at}_{mk} and S'^{at}_{km} are similar processes occurring upon atom-atom collisions, and S_k is the rate of ionization by electron impact from the kth state. The quantities S_{mk}, S_{km}, and S_k are functions of the electron temperature T_e, whereas S'^{at}_{mk} and S'^{at}_{km} are those of the gas temperature T_g. A'_{mk} and A'_{kl} are the effective probabilities that atoms will decay spontaneously in a radiative fashion from the mth to the kth level and from the kth to the lth level, respectively, with due regard for radiation trapping and τ_m is the time metastable atoms take to reach the discharge tube walls by diffusion. The last term in expression (2.11) allows for the stimulated radiative transitions $k \rightarrow m$ under the effect of the external resonant radiation with a photon flux of F and a photoexcitation cross section of σ_{ph}, the symbols g_m and g_k standing for the statistical weights of the mth and kth levels, respectively.

In writing down expression (2.11), it is assumed that the density of atoms in the ground state substantially exceeds the density of the excited atoms: $n_1 \cong n_a >> \sum\limits_{k=1}^{r} n_k$, where n_a is the total atomic density. Ionization resulting from atom-atom collisions is not taken into account.

The role of the radiation trapping effect is taken into consideration by a simple procedure consisting in the multiplication of the ordinary radiative transition probabilities into the appropriate trapping factor γ [2.3]. Generally speaking, this factor depends on the geometry of the object and the type of broadening of the absorption and emission lines. However, for approximate estimation purposes, it frequently proves sufficient to take simple expressions for the factor γ. To illustrate, for a long cylinder of radius R and a dispersion line profile,

$$\gamma = \left(2\sqrt{k_0 R}\right)^{-1}, \tag{2.12}$$

where k_0 is the absorption coefficient at the line center.

For a Doppler line profile,

$$\gamma = \sqrt{\pi}\left(4 k_0 R \sqrt{\ln k_0 R}\right)^{-1}. \tag{2.13}$$

The time it takes for metastable atoms to reach the cylinder walls by diffusion is given by the expression

$$\tau_m^{-1} = D_m / \Lambda^2, \quad \Lambda = R / 2.4, \tag{2.14}$$

where D_m is the diffusion coefficient of the atoms.

The electron temperature T_e is found from the energy balance equation:

$$(3/2)k\frac{dT_e}{dt} = \frac{e^2 E^2}{m_e \nu_t} + W_{se} - W_{el} - W_{ex} - W_i, \tag{2.15}$$

where E is the axial electric field strength in the positive column, e and m_e are the electronic charge and mass, respectively, ν_t is the transport electron collision frequency, k is the Boltzmann constant, and the quantities W_{se}, W_{el}, W_{ex}, and W_i define the contributions from superelastic, elastic, exciting, and ionizing collisions, respectively.

The charged particle balance is defined by the equation

$$\frac{dn_e}{dt} \equiv G = n_e \sum_{i=1}^{r} n_i S_i + D^+ \nabla^2 n_e, \qquad (2.16)$$

where D^+ is the ambipolar diffusion coefficient.

The total density of particles in the tube volume is conserved:

$$n_e + \sum_{i=1}^{r} n_i = n_a. \qquad (2.17)$$

The electric field strength E, the current i, and the conductivity $\sigma(T_e, n_e)$ in the positive column of a glow discharge are related together by Ohm's law for a subcircuit [2.2]:

$$E = i / \sigma(T_e, n_e), \qquad (2.18)$$

where $\sigma(T_e, n_e) = 1.36e\, \mu_e n_e(0)R^2$, μ_e is the electron mobility, and $n_e(0)$ is the electron density on the axis of the discharge tube.

The resonant absorption of light from an external source, which gives rise to the optogalvanic effect and enters into expression (2.11), supplements, in this case, the well-known positive column models [2.2]. It should be noted that the outward simplicity of the system of equations (2.11), (2.15)–(2.18) hides complex nonlinear relationships between the quantities G, K, σ, and W on the one hand and a great many variables on the other. What is more, the number of balance equations for expression (2.11) may prove fairly great in some or other concrete situation.

The mathematical difficulties involved in solving the above system of equations are due to the their property of being *stiff*, which imposes special requirements upon the numerical algorithm used for the purpose. This is explained by the fact that the transitions between the ground state and the excited states occur on a time scale much longer than that of the relaxation processes associated with the transitions between the levels being excited or between the ionization continuum and the bound excited states. To solve such systems of stiff ordinary differential equations numerically, it is expedient to use the algorithm suggested by Geer (see [2.4]).

In addition to the numerical methods of solving the given problem, an analytical approach has been suggested [2.5]. This approach can be used when the discharge current variation due to the optogalvanic effect is small, so that the magnitude of the effect can be taken to be linear in the number of photons absorbed. However, when using this method, the resultant model proves to be open, semiempirical.

Thus, the general solution of the problem is split into solutions of two subproblems, the first being equations of the radiative-collisional model, which ultimately define the electric field strength variation ΔE in the positive column. Next the response Δi of the discharge current is found to suit the given type of discharge.

2.2 EFFECT OF IRRADIATION ON POSITIVE COLUMN CHARACTERISTICS

The magnitude of the steady-state optogalvanic effect in the positive column of a discharge was first calculated in [2.6]. In this work, a three-level sodium atom model ($r = 2$) was analyzed on the basis of Eqs. (2.11) and (2.16). Account was taken of the ground, resonance, and ionization states (Fig. 2.1). The population of the excited state was governed by the laser radiation, electron impact, and spontaneous decay effects. The gas was ionized by electron impact from the ground and excited states. The loss of charged particles was determined by their ambipolar diffusion onto the positive column wall. The energy distribution of electrons was taken to be Maxwellian. Therefore, based on the detailed balancing principle, the atomic electron-impact excitation and de-excitation coefficients were related together by the relation

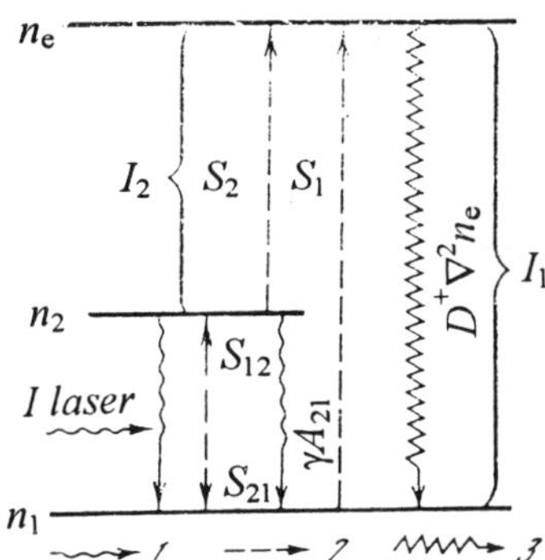

Figure 2.1. Atomic energy level and transition diagram for the model analyzed in [2.6]. I_1 and I_2 – ionization potentials; *1* – optical transition; *2* – electron collisions; *3* – ambipolar diffusion.

$$S_{mk} = \frac{g_m}{g_k} S_{km} \exp\left[-E_{km} / kT_e\right], \quad m > k, \tag{2.19}$$

where E_{km} is the energy interval separating the levels k and m. The analysis took into account the buffer gas that reduced the rate at which charged particles diffused onto the discharge tube walls and lowered the electron temperature accordingly. Also included in the balance equations for the atoms in the ground state was the term $(-D^+\nabla^2 n_e)$ describing their diffusion away from the walls at which they were formed as a result of the electron-ion recombination process. It was assumed, too, that the electron concentration suffered no changes as a result of the optogalvanic effect.

In energy balance equation (2.15), account was taken of the energy loss due only to the ionization of the gas, so that it assumed the form [2.6]

$$Ei = 7.85 n_e D^+ \left(I_1 + 5kT_e\right), \tag{2.20}$$

where I_1 is the ground state ionization potential.

The solution of Eq. (2.16) for a cylindrical charge configuration is a Bessel function of order zero. Considering the fact that the electron concentration on the discharge tube walls must be equal to zero, the following expression holds true[2.2]:

$$\sum_{i=1,2} n_i S_i = D^+ / \Lambda^2. \tag{2.21}$$

In that case, one can find from Eqs. (2.20) and (2.21) the magnitude of the optogalvanic response as the ratio $\Delta E/E$, provided that the discharge current remains constant:

$$\frac{\Delta E}{E} = \frac{\sum\limits_i \left[\left(I_1 + 5kT_e^0\right)\Delta\left(n_i S_i\right) + 5k\left(n_i S_i\right)^0 \Delta T_e\right]}{\sum\limits_i \left(I_1 + 5kT_e^0\right)\left(n_i S_i\right)^0}. \tag{2.22}$$

The quantities denoted by the superscript 0 are calculated in the absence of the external radiation.

The calculations were made for $R = 0.5$ cm, a buffer gas pressure of $p = 10$ Torr, and $n_e = 10^{13}$ cm^{-3}. Figure 2.2 presents the quantity $-\Delta E/E$ as a

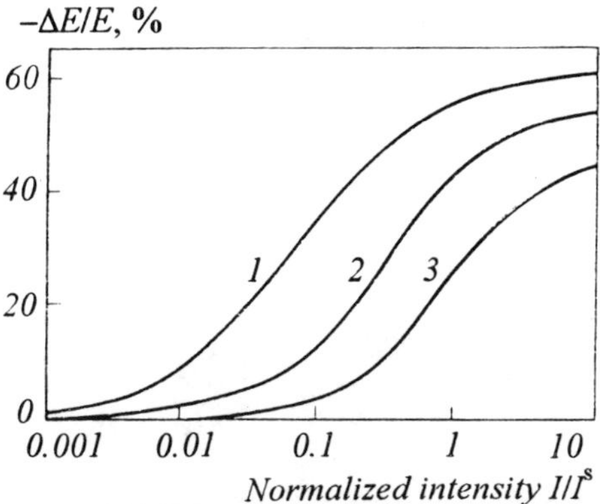

Fig. 2.2. Optogalvanic response as a function of the external radiation intensity and saturated sodium vapor temperature [2.6]. $1 - T = 800$ K, $I^s = 1.9$ mW·cm^{-2}; $2 - T = 600$ K, $I^s = 2.83$ mW·cm^{-2}; $3 - T = 500$ K, $I^s = 6.79$ mW·cm^{-2}.

function of the normalized radiation intensity at various temperatures of the saturated sodium vapor used. The symbol I^s in the figure stands for the saturating radiation intensity. The radiant excitation of the positive columns was taken to be uniform.

The sign of the optogalvanic response is negative. This is due to the fact that irradiation increases the population of the excited state, and this in turn reduces the electron temperature (see Eq. (2.21)) and thus lowers the electric field strength accordingly (see Eq. (2.20)).

The drop of I^s with the rise of the vapor temperature is due to the lowering of the probability of spontaneous decay as a result of radiation trapping.

The numerical results presented in Fig. 2.2 show that the lowering of E may come to a few tens percent.

This model was later modified in [2.7]. The authors took account of the fact that the variation of the electric field strength at a constant discharge current was associated with both the change of the temperature of the electrons and the change of their concentration. The following value was used in the calculations for the electron mobility μ_e:

$$\mu_e = (3/4)(3/8\pi)^{1/2} / e\lambda_e / \left[m_e (3kT_e / m_e)^{1/2} \right], \tag{2.23}$$

where λ_e is the electron free path length. Accordingly, the expression for the discharge current will have the form

$$i = 0.54 n_e e^2 R^2 \lambda_e E / \left(2 m_e k T_e / \pi\right)^{1/2}. \qquad (2.24)$$

In that case, we have from equation (2.24) the following expression for ΔE at $i = \text{const}$:

$$\Delta E = 1.96 \frac{\left(2 m_e k T_e^0 / \pi\right)^{1/2}}{e^2 R^2 \lambda_e n_e^0} \left(\frac{\Delta T_e}{2 T_e^0} - \frac{\Delta n_e}{n_e^0}\right) i. \qquad (2.25)$$

The authors of [2.7] conducted their experiment in a discharge positive column of length $L = 4$ cm and radius $R = 0.15$ cm. The calculation results and experimental data are presented in Fig. 2.3 for a discharge current of $i = 5$ mA, a sodium atom concentration of $n_1 = 4.35 \times 10^9$ cm^{-3}, and a buffer gas pressure of $p = 50$ Torr. One can clearly see that the quantities Δn_e, ΔT_e, and ΔE get saturated as the intensity of the external laser radiation used is raised.

It should be noted that under the experimental conditions used in [2.7] both terms in the parentheses of Eq. (2.25) are commensurable, being of

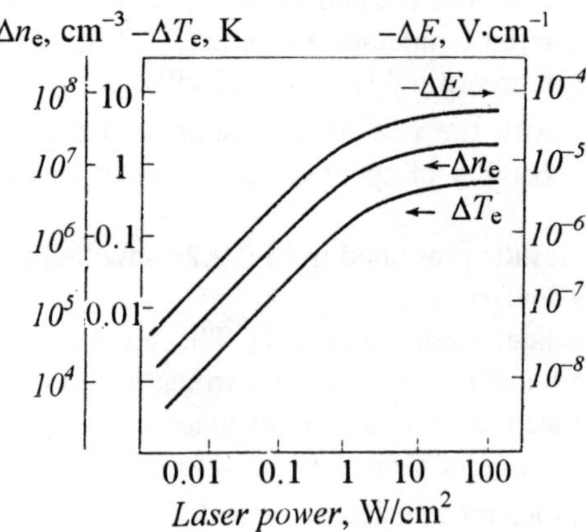

Fig. 2.3. Δn_e, Δt_e, and ΔE as a function of the laser radiation intensity [2.7].

the order of 10^{-4}, and so the change of the electron concentration has a great effect on the magnitude of the optogalvanic response.

One should also mention that no consideration was given in [2.6] and [2.7] for such important factors as the superelastic heating of the electrons and the loss of energy through emission. And whereas their role in [2.7] is insignificant, the sodium vapor pressures used being low, these factors are important in conditions used in [2.6], as will be shown later.

More detailed calculations of the characteristics of the positive column in a resonant electromagnetic field were performed in [1.4].   At the first stage, the qualitative behavior of the optogalvanic effect was analyzed on the basis of the above-mentioned three-level model. When the discharge current is small and the external radiation intensity is weak, the excited state is populated by the resonant radiation, and its depletion occurs as a result of an effective radiative decay. In that case,

$$n_2 = F\sigma_{\mathrm{ph}} / \left(A_{21}\gamma \right). \tag{2.26}$$

To determine the electron temperature, use was made of Eq. (2.21). Considering expression (2.26), we have

$$S_1 n_1 + S_2 F\sigma_{\mathrm{ph}} n_1 / \left(A_{21}\gamma \right) = D^+ / \Lambda^2 . \tag{2.27}$$

Let us represent the electronic ionization cross section of the atom in the state i in a form convenient for calculation and used frequently in the literature [2.2]:

$$\sigma_i = c_i\left(\varepsilon_i - I_i \right), \tag{2.28}$$

where ε_i is the electron energy. In that case, for a Maxwellian electron energy distribution, the ionization coefficient is given by

$$S_i \cong c_i I_i \left(8kT_e / \pi m_e \right)^{1/2} \exp\left(-I_i / kT_e \right). \tag{2.29}$$

Let us also write down the expression for the ambipolar diffusion coefficient [2.2]:

$$D^+ = kT_e\mu_i / e, \tag{2.30}$$

where μ_i is the ion mobility.

Substituting expressions (2.29) and (2.30) into Eq. (2.27), we finally obtain the following relation:

$$\frac{\exp\left(I_1 / kT_e\right)}{\left(I_1 / kT_e\right)^{1/2}\left\{1+\left[B_{12}I_1 C_2 I_2 /\left(A_{21} C_1 I_1\right)\right]\exp\left[\left(I_1 - I_2\right)/kT_e\right]\right\}} \tag{2.31}$$
$$= 1.2 \times 10^7 (cR)^2 p_1 p_2,$$

where p_1 and p_2 are the pressures of the resonant and buffer gases, respectively, and the constant c is defined by the expression

$$c = \left[3.6 \times 10^{16} C_1 I_1^{1/2}\left(\mu_i p_2\right)\right]^{1/2}. \tag{2.32}$$

Specifically, for sodium vapor in neon, $c = 10^{-2}$.

Note also that the estimates presented in [1.4] have shown that at a pressure of $p_2 = 1$ Torr the contribution of the buffer gas to ionization can be disregarded, provided that $p_1 >> 10^{-3}$ Torr.

The solution of transcendental equation (2.31) allows one to find the electron temperature T_e as a function on the concentration of the gases and the external radiation intensity. Irradiation causes an increase in the population of the excited state and raises the ionization probability of the gas. According to (2.27), the total number of ionization events should practically remain unchanged, and so T_e will be reduced with the increasing intensity of the resonant light field.

Thereafter, the axial electric field strength in the positive column was calculated with due regard for the effect of the resonant electromagnetic field. To this end use was made of energy balance equation (2.15) for an electron system in a three-level gas. Subject to analysis were both the resonant and buffer gases. The electron energy balance allowed for the losses through the heating, excitation, and ionization of the gases, as well as for the heating of the electrons upon superelastic collisions with the excited atoms.

The effect of the additional resonant field gives rise to two consequences affecting the electric field strength E. The first is associated with the change in the electron temperature governing the electronic coefficients S and the second is due to the change in the excited state population n_2 and hence the role of superelastic collisions in the electronic energy balance. Obviously the rise of the radiation intensity reduces the electron temperature T_e and increases the population n_2, which on the whole lowers the electric field strength E.

Figure 2.4 presents concrete calculation results for the characteristics of a positive column with a radius of $R = 0.5$ cm containing neon as a buffer gas at a pressure of $p_1 = 1$ Torr with an addition of sodium vapor excited by an external radiation with a wavelength of λ 589 nm.

The effect of the resonant radiation starts manifesting itself at a photon flux of $F \gg 10^{11}$ photons/cm^2·s when the stepwise electronic ionization of sodium atoms comes into play. As the resonant gas (sodium vapor) pressure and radiation power are increased, the electron temperature drops. This can easily be understood by analyzing expression (2.27) wherein the right-hand side varies but little, and so the quantities S_1 and S_2 must decrease to offset the increase of n_1 and F.

Presented in Fig. 2.5 are the most important partial contributions from the individual processes to the energy balance. As can be seen, the main electron energy loss is associated with the excitation of the sodium vapor. At a pressure of $p_1 > 5 \times 10^{-3}$ Torr the contribution of the buffer gas to ionization is small. The superelastic processes can grow in importance in the energy balance as the pressure of the resonant gas is raised.

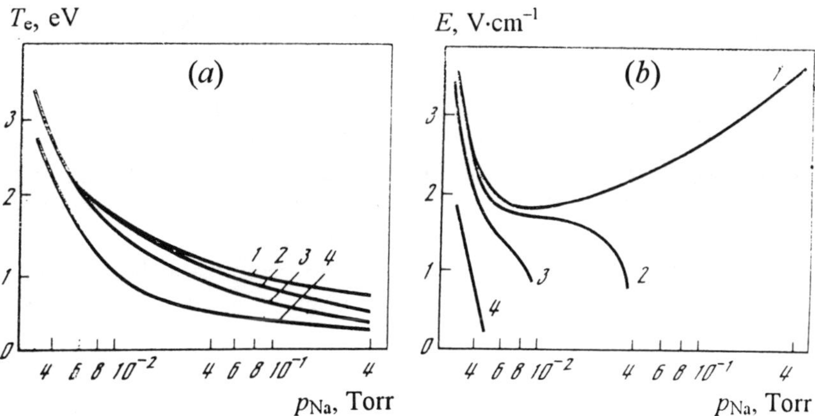

Figure 2.4. (a) Electron temperature in the positive column [1.4] and (b) axial electric field strength in the positive column [1.4] as a function of the Na vapor pressure. Photon flux F (in photons/cm^2·s): $1 - 0$; $2 - 7 \times 10^{12}$; $3 - 7 \times 10^{13}$; $4 - 7 \times 10^{14}$.

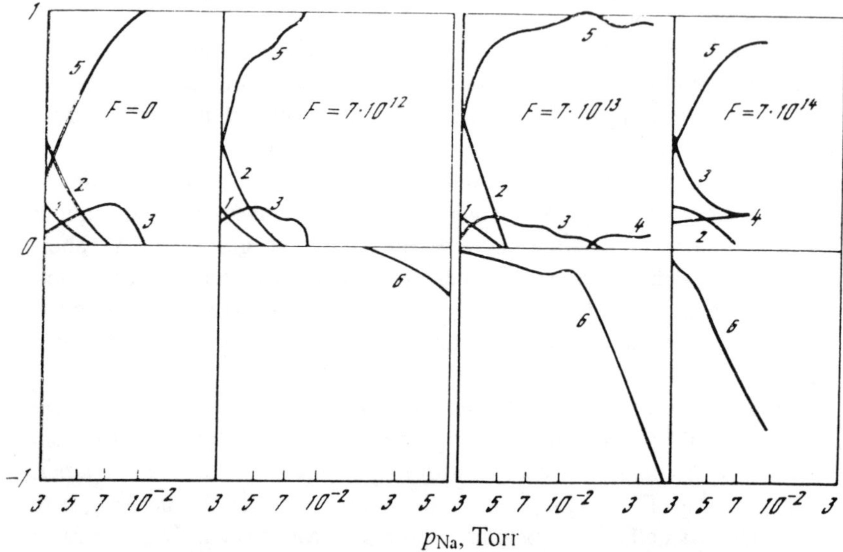

Figure 2.5. Partial losses of various processes and their contribution to the energy balance [1.4]. *1* and *2* – ionization and excitation of ground-state Ne atoms, respectively; *3* and *4* – ionization of ground-state and excited Na atoms, respectively; *5* – excitation of the resonant state in Na; *6* – superelastic collisions with the excited resonant Na atoms.

Of great interest are studies into the optogalvanic effect in the discharge plasmas of inert gases, neon in particular. This is due, on the one hand, to the widespread occurrence of such objects, and on the other, to the fact that a great many transitions in neon fall within the lasing range of lasers using Rhodamine 6G dye as their active medium (Fig. 2.6).

When analyzing the optogalvanic effect in neon, the authors of [2.8] replaced the diffusion term in electron balance equation (2.16) by the expression n_e/τ_a, where

$$\tau_a^{-1} = D^+ / \Lambda^2 \qquad (2.33)$$

is the characteristic time it takes for metastable neon atoms to reach the discharge tube walls by diffusion, defined by expression (2.14).

To calculate the optogalvanic signals numerically, is was assumed that the heating of the gas in the discharge tube was insignificant, so that $T_g =$

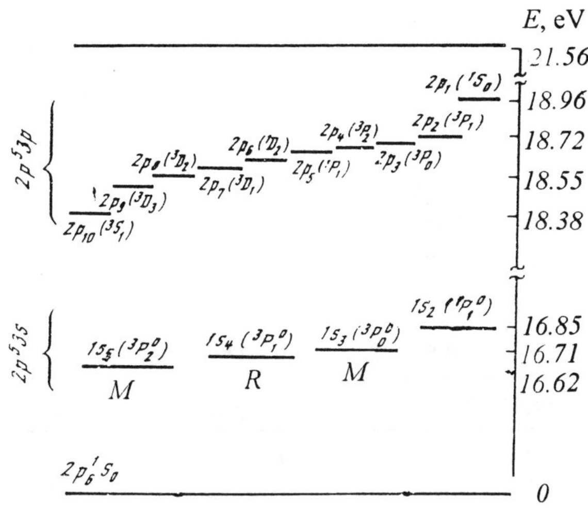

Figure 2.6. Diagram of the energy levels considered in the Ne atom.

const $= 300$ K, $D^+ \cong 3.5{\times}10^3 T_e/p$ (in $cm^2{\cdot}s^{-1}$), T_e is in electron-volts, p is the neon pressure in torrs, and $D_m n_1 = 5.3{\times}10^{18}$ $cm^{-1}{\cdot}s^{-1}$.

The cross section of elastic collisions between electrons 1.5 to 5.1 eV in energy and neon atoms in the ground state is $\sigma_{el} \cong 1.9{\times}10^{-16}$ cm^2, $\sigma_{el} \cong \sigma_t$, where σ_t is the transport collision cross section.

Taken into consideration in the neon atom model were the ground and ionization states, along with a block of the lower $1s$ (4 levels) and upper $2p$ (10 levels) excited states (see Fig. 2.6). Two levels in the lower block, namely, $1s_2$ and $1s_4$, are resonant, and the other two, $1s_3$ and $1s_5$, are metastable. The levels $2p$ are optically coupled to their $1s_i$ counterparts. One of the transitions, namely, $1s_i \to 2p_j$ is acted upon by an external resonant radiation which uniformly excites the positive column of the discharge.

The rate S of the electron-impact mixing of the levels $1s$ in the lower block was taken to be the same for all levels and equal to $4{\times}10^{-7}$ $cm^3{\cdot}s^{-1}$ [2.9]. The rates of direct atomic collisions mixing the levels in the blocks $1s$ and $2p$ are listed in Tables 2.1 and 2.2 in accordance with the data presented in [2.10]. The rates of the reverse processes were calculated on the basis of the detailed balancing principle.

Table 2.1. Atomic-collision mixing rates for levels $1s$, S_{ij}^{at} ($\times 10^{-16}$ cm$^3 \cdot$s^{-1}) [2.10].

i	$1s_5$	$1s_5$	$1s_5$	$1s_4$	$1s_4$	$1s_3$
j	$1s_4$	$1s_3$	$1s_2$	$1s_3$	$1s_2$	$1s_2$
S_{ij}^{at}	34	0.3	0.009	3	10	2

Table 2.2. Atomic-collision mixing rates for levels $2p$, S_{ij}^{at} ($\times 10^{-13}$ cm$^3 \cdot$s^{-1}) [2.10].

	$2p_1$	$2p_2$	$2p_3$	$2p_4$	$2p_5$	$2p_6$	$2p_7$	$2p_8$	$2p_9$	$2p_{10}$
$2p_1$	–	0	0	0	0	0	0	0	0	0
$2p_2$	0	–	7	21	7	5	4	2	0	1
$2p_3$	0	7	–	2100	41	0	10	5	15	10
$2p_4$	0	6	320	–	330	22	54	5	4	0
$2p_5$	0	1	5	350	–	7	49	4	0	2
$2p_6$	0	0	3	1	4	–	109	36	7	4
$2p_7$	0	0	0	0	0	80	–	72	17	12
$2p_8$	0	0	0	0	0	80	16	–	23	80
$2p_9$	0	0	0	0	0	10	6	130	–	420
$2p_{10}$	0	0	0	0	0	0	0	0	0	–

The electron-impact excitation rates for the atoms were calculated on the basis of the formulas given in [2.11].

No data were presented in [2.10] on the rates of mixing of the levels $2p_1$ with the levels $2p_i$ ($i \neq 0$). Considering the high energies of the corresponding transitions, these rates were taken to be zero.

The calculations were performed for the case of radiative excitation of the transitions $1s_5 \to 2p_8$, $1s_5 \to 2p_4$, $1s_4 \to 2p_8$, $1s_5 \to 2p_2$, and $1s_2 \to 2p_8$.

When numerically calculating the relationship between the magnitude of the optogalvanic effect and the discharge current, the flux density of the external resonant radiation was taken to be $F = 1.18 \times 10^{16}$ photons/cm$^2 \cdot$s, which corresponded to a radiation power of 4 mW (for the transition $1s_5 \to 2p_2$, $\lambda = 5855$ Å), i. e., the radiation power was far from the saturation value ($F^n \sim 6 \times 10^{17}$ photons/cm$^2 \cdot$s, see below).

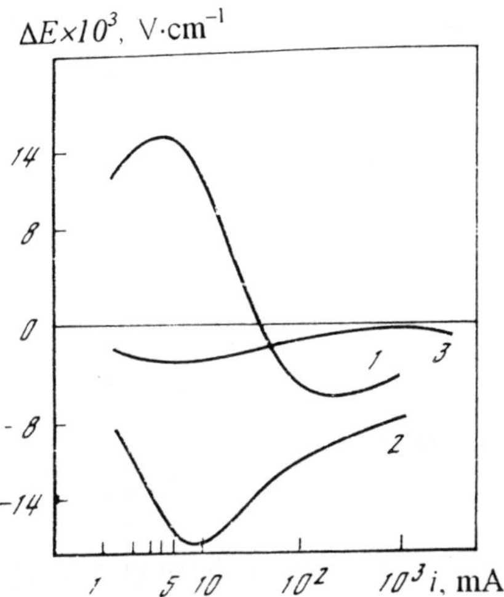

Figure 2.7. ΔE as a function of the discharge current i ($p = 1$ Torr, $R = 4$ mm), mA. *1* – transition $1s_5 \rightarrow 2p_8$ ($\lambda = 6334.4$ Å); *2* – transition $1s_4 \rightarrow 2p_8$ ($\lambda = 6506.5$ Å); *3* – transition $1s_2 \rightarrow 2p_8$ ($\lambda = 7173.9$ Å).

The calculation results for the electric field variation in the positive column are presented in Fig. 2.7 (gas pressure 1 Torr, discharge tube radius 4 mm). The optogalvanic responses to the resonance states vary in amplitude without changing sign. If the transitions subject to excitation start with metastable states, the optogalvanic signal changes sign as the discharge current is increased to 40–60 mA. When the discharge current is further increased, the optogalvanic responses for both types of transition behave similarly and do not change sign, their amplitudes tending to zero while remaining negative.

The reasons for such behavior of the optogalvanic signals are as follows. The lower block of excited states consists of metastable levels $1s_m$ ($1s_5$ and $1s_3$) and resonance levels $1s_r$ ($1s_4$ and $2s_2$) that are optically coupled to the ground state s_0. The levels of the block $2p$ are optically coupled to all the levels $1s_i$. The action of radiation on the transition $1s_i \rightarrow 2p_j$ may lead to the realization of one of the following cases:

$$
\begin{aligned}
&\text{(a)} && 1s_{\mathrm{m}} \to 2p_j - \Big|
\begin{array}{l}
\to 1s_{\mathrm{r}} \ (> 50\%) \\[2pt]
\to 1s_{\mathrm{m}} \ (< 50\%),
\end{array} \\[8pt]
&\text{(b)} && 1s_{\mathrm{m}} \to 2p_j - \Big|
\begin{array}{l}
\to 1s_{\mathrm{m}} \ (> 50\%) \\[2pt]
\to 1s_{\mathrm{r}} \ (< 50\%),
\end{array} && (2.34) \\[8pt]
&\text{(c)} && 1s_{\mathrm{r}} \to 2p_j - \Big|
\begin{array}{l}
\to 1s_{\mathrm{m}} \ (> 50\%) \\[2pt]
\to 1s_{\mathrm{r}} \ (< 50\%),
\end{array} \\[8pt]
&\text{(d)} && 1s_{\mathrm{r}} \to 2p_j - \Big|
\begin{array}{l}
\to 1s_{\mathrm{r}} \ (> 50\%) \\[2pt]
\to 1s_{\mathrm{m}} \ (< 50\%).
\end{array}
\end{aligned}
$$

The figures $> 50\%$ or $< 50\%$ indicate the proportion of particles finding themselves at the given type of level after the spontaneous decay of the state $2p_j$. The formation of the optogalvanic signal depends on several factors:

(1) population redistribution among the metastable and resonance levels of the lower block following the spontaneous decay of the state $2p_j$ in accordance with scheme (2.34);

(2) increase of the ionization rate under the effect of the external radiation (on account of the population of the $2p_j$ state);

(3) change of the effective ionization potential as a result of the population of the higher-lying levels of the lower block following the spontaneous decay of the state $2p_j$.

Analysis of the numerical calculation results shows that the first factor plays the most important role. In cases (a) and (b) (scheme (2.34)), there takes place the depletion of the entire $1s$ block (via the resonance states), case (a) being more effective. Thus, in cases (a) and (b), the number of excited atoms on the whole diminishes causing T_{e} and E to rise under the effect of the external radiation.

In cases (c) and (d), irradiation results in the overpopulation of the block $1s$, which leads to an increase in the number of excited atoms, and ΔT_{e} and ΔE are in that case negative.

Increasing the discharge current, hence n_{e}, enhances the electron-impact mixing of the levels in the lower block, so that at currents of 40–60 mA the spontaneous decay rate of the resonance states ($1s_2$, $1s_4$), with due regard for radiation trapping, becomes comparable with the electronic mixing rate of the levels in the lower block. In that case, all the $1s_i$ levels can be treated as a single quasilevel with a lifetime close to that of the resonance state. The action of the external radiation here boils down to the additional excitation of the $2p_j$ state more advantageous for ionization, and so the field in that case decreases, no matter what the type (resonance or metastable) of the state acted upon by the radiation. As the discharge current is further increased, the optogalvanic signal amplitude (ΔE in the given case) diminishes, which is associated with a decrease of the role of the external radiative processes in comparison with their electronic counterparts.

Figure 2.8 illustrates the variation of the electron temperature under the effect of an external radiation on various transitions as a function of the discharge current. As the current is increased, the influence of the external radiative processes on the discharge decreases. At small discharge currents, the behavior of ΔT_e is governed by the population redistribution among the levels of the block $1s$, and at discharge currents in excess of 60 mA, by the

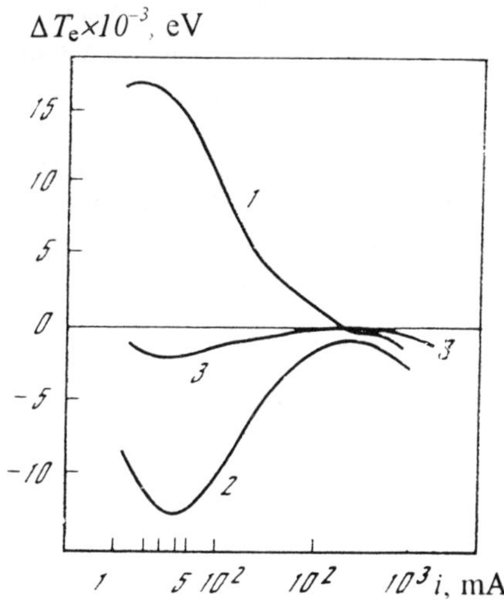

Figure 2.8. ΔT_e as a function of the discharge current i ($p = 1$ Torr, $R = 4$ mm), mA. 1 – transition $1s_5 \rightarrow 2p_8$ ($\lambda = 6334.4$ Å); 2 – transition $1s_4 \rightarrow 2p_8$ ($\lambda = 6506.5$ Å); 3 – transition $1s_2 \rightarrow 2p_8$ ($\lambda = 7173.9$ Å).

drop of the ionization potential upon transition to the $2p_j$ state. The electron temperature variation ΔT_e in that case decreases in amplitude with the increasing discharge current for all the transitions while always remaining negative.

As already noted, the role of inelastic processes in transitions $1s \rightarrow 2p$ is small, the decisive part being played by the decrease of the ionization potential under the effect of the external radiation. In that case, as the discharge current is increased the amplitude of ΔT_e tends to zero, for the relative contribution from the radiative processes is reduced.

2.3 CHANGES IN NORMAL GLOW DISCHARGE CHARACTERISTICS UNDER THE EFFECT OF RADIATION

The author of [2.5] developed a simple linear model for a stationary optogalvanic effect in a helium discharge exposed to radiation with a wavelength of $\lambda = 587.\,6$ nm in conditions far from saturation.

Analyzed in this work was the $2^3P \rightarrow 3^3D$ transition with well-known elementary process cross sections. The electronic ionization of both levels was disregarded. The dominant process determining the decay of the 3^3D state is the spontaneous transition to the 2^3P state. The associative ionization process taking place when a normal atom collides with an excited one gives rise to a molecular helium ion and an electron. In this situation, the efficiency of ionization caused by the absorption of photons is determined by the ratio between the associative dissociation rate a and the spontaneous decay rate A.

The change in the optogalvanic signal current amounts to a few fractions of a percent. This gives reason to believe that the relationship between Δi and the number of photons absorbed is linear. Experimental results show that this condition is only disturbed when the resonance transition absorption is saturated. The model suggested by the author of [2.5] fails to give the relationship between the optogalvanic effect and the spatial distribution of the photons absorbed. However, the results of the experiment show that when the radiation is focused on the discharge axis the optogalvanic response is greater by 10–20% compared to that in the case of uniform distribution of the radiation.

The main unknowns in the above model are the electron concentration n_e on the discharge axis and the axial electric field E. The positive column radius R and the gas pressure p are fixed.

At the root of the mathematical model are kinetic equation (2.16) for ions, which defines their creation and annihilation,

$$G(n_e, E) = 0, \qquad (2.35)$$

and the equation for the discharge current

$$i = F(n_e, E) = \sigma E. \qquad (2.36)$$

The magnitude of the current i is specified by the external circuit conditions, and the variation of the electron temperature T_e is neglected. Exposing the discharge to laser radiation results in the perturbation of E and n_e, defined by the equations

$$\frac{\partial G}{\partial n_e} \Delta n_e + \frac{\partial G}{\partial E} \Delta E + \frac{a}{A} Q = 0 \qquad (2.37)$$

and

$$\Delta i = \frac{\partial F}{\partial n_e} \Delta n_e + \frac{\partial F}{\partial E} \Delta E, \qquad (2.38)$$

where a/A is the ionization efficiency and Q is the number of photons absorbed per unit time.

Finding ΔE and Δn_e from Eqs. (2.37) and (2.38) and substituting the result into Eq. (2.5), we get

$$\Delta i = \frac{a}{A} Q \frac{\partial F}{\partial n_e} \frac{L}{z} \left[\frac{\partial G}{\partial n_e} \left(\frac{\partial F}{\partial E} + \frac{L}{z} \right) - \frac{\partial G}{\partial E} \frac{\partial F}{\partial n_e} \right]. \qquad (2.39)$$

Using Eqs. (2.35) and (2.36), one can obtain the total derivative of the discharge current with respect to the electric field strength:

$$\frac{di}{dE} = \frac{\partial F}{\partial E} - \frac{\partial F}{\partial n_e} \frac{\partial G}{\partial E}. \qquad (2.40)$$

Introducing the dynamic resistance of the discharge,

$$\frac{dV}{di} = L \bigg/ \left(\frac{di}{dE}\right), \quad LdE = dV, \tag{2.41}$$

and taking into account Eqs.(2.40) and (2.41), one may rearrange Eq.(2.39) in the form

$$\Delta i = -\frac{a}{A}Q\left(\frac{\partial F}{\partial n_e}\bigg/\frac{\partial G}{\partial n_e}\right)\frac{dV}{di}\left(\frac{dV}{di}+z\right)^{-1}. \tag{2.42}$$

The magnitude of the dynamic resistance is measured experimentally.

The positive column was analyzed in free fall conditions on the basis of the Tonks-Langmuir model [2.2] under the following assumptions.

1. The loss of the metastable atoms excited by electron impact results from their diffusion onto the discharge tube wall, and so their density is proportional to the electron density.

2. The ionization of atoms takes place in (a) a two-step electron-atom interaction process proceeding via a metastable state and (b) an associative interaction process involving an atom in a highly excited state populated by electrons from a metastable state and an atom in the ground state, giving rise to a molecular ion. The ionization rate is in that case proportional to the electron concentration squared, n_e^2.

3. The recombination of the ions occurs on the wall.

The magnitude of the discharge current is governed by the electron drift:

$$i = F(n_e, E) = en_e\mu_e E\pi R^2 2h_0, \tag{2.43}$$

where h_0 is the Tonks-Langmuir constant arising in the averaging of the electron concentration over the discharge tube radius [2.2].

The kinetic equation defining the electron concentration has the form

$$G(n_e, E) = g(E)n_e^2 - 2(2kT_e/M_i)^{1/2}2\pi RLS_0h_0n_e = 0, \tag{2.44}$$

where $g(E)$ depends on the electron impact rate depending on E/p, metastable atom diffusion coefficient, elastic collision rate, and the parameters R and p, M_i is the ionic mass, and $S_0 = 0.77$ is a constant in the Tonks-Langmuir theory [2.2].

Based on Eqs. (2.43) and (2.44), we get

$$\frac{\partial F}{\partial n_e}\bigg/\frac{\partial G}{\partial n_e} = eR\mu_e E\left[S_0 L\left(\frac{2kT_e}{M_i}\right)^{1/2}\right]^{-1}. \qquad (2.45)$$

Substituting Eq. (2.45) into Eq. (2.42), we obtain the following expression for the optogalvanic response:

$$\Delta i = -\frac{a}{A}Q\left\{eR\mu_e E\left[0.77L\left(\frac{2kT_e}{M_i}\right)^{1/2}\right]^{-1}\right\}\frac{dV}{di}\left(\frac{dV}{di}+z\right)^{-1}. \qquad (2.46)$$

The experiment was conducted with discharge tubes differing in diameter. Use was made of a single-frequency tunable laser 1 Mhz in spectral bandwidth, much narrower than the Doppler width equal to 2 GHz. Spatially homogeneous laser radiation was directed along the discharge axis. Subject to measurement was the radiation power absorbed in the discharge, the voltage drop across the ballast resistor, and the dynamic resistance of the discharge. Despite the fact that the laser radiation was modulated with a frequency of 90 Hz, the use of the stationary model was quite justified, for the relaxation time in the discharge was much shorter than the modulation period.

Figure 2.9 presents the values of (a) the dynamic resistance and (b) current response for $Rp = 0.05$ cm·Torr. The following values were used in calculations: $a = 2.4\times10^6 p$ s^{-1} (p is measured in torrs), $A = 7.06\times10^7$ s^{-1}, s$\mu_e = 7.5\times10^5/p$ cm^2·(V·s)$^{-1}$, and $L = 13.5$ cm.

The magnitude of the axial electric field strength E was measured for each Rp value. The T_e values were borrowed from [2.12] for various Rp values. The discharge current dependences of E/p and T_e were not taken into consideration. The quantity M_i was taken to be the mass of the molecular helium ion resulting from the associative ionization process.

A good agreement was achieved in the above work between theoretical and experimental response values for $0.05 \leq Rp \leq 0.5$ cm·Torr.

This approach was later extended to the study of the optogalvanic effect in neon at a wavelength of $\lambda = 594.5$ nm [2.13]. The action of radiation leads to the destruction of metastable states, reduction of the conductivity of the positive column plasma, and hence the formation of a negative optogalvanic response.

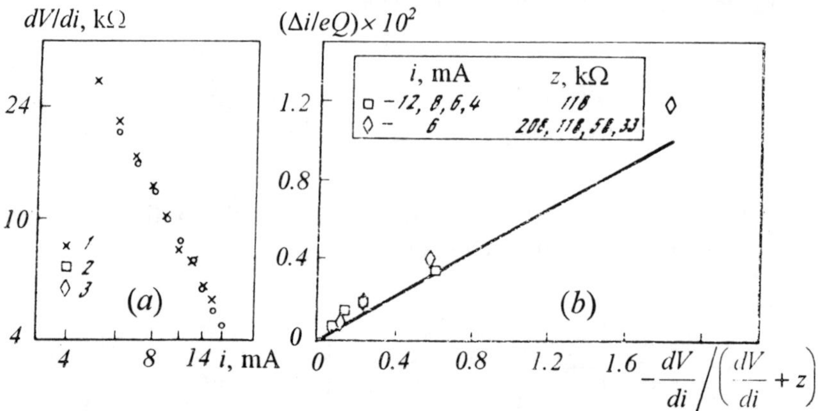

Figure 2.9. (*a*) Dynamic resistance as a function of the discharge current and (*b*) discharge current as a function of the quantity $-(dV/di)/(dV/di + z)$ [2.5]. (R = 0.1 cm, p = 0.52 Torr, E = 39–45 V/cm). *1* – measurement results for the entire positive column; *2* – optogalvanic response at a fixed ballast resistance and fixed discharge current values; *3* – optogalvanic response at a fixed discharge current and different ballast resistance values. The solid line represents theoretical calculation on the basis of the model suggested.

The mathematical model describing the optogalvanic effect consists of equations for metastable atoms, electrons, and the discharge current.

It is the authors' opinion that the group of levels $1s_5$, $1s_3$, and $1s_4$ can be united into a single quasilevel M, for they all have a long lifetime and are strongly coupled together by electron impact (see Fig. 2.6). The authors disregard the influence of the level $1s_2$ on ionization, assuming that its lifetime is too short.

The population n_m of the level M is defined by the equation

$$\frac{dn_m}{dt} = H(n_e, n_m\, E) = \widetilde{S}n_e n_1 - \widetilde{A}n_m - Tn_m^2 - S_m n_e n_m$$

$$-\frac{S'n_e n_m}{4} = 0, \tag{2.47}$$

where $\widetilde{S}$ is the rate of electronic excitation of the metastable state both directly from the ground state and indirectly in the course of radiative decay of the higher-lying levels populated from the ground state by electron impact. The magnitude of $\widetilde{S}$ strongly depends on E/n_1. The quantity $\tilde{A}$ determines the loss of atoms as a result of spontaneous process, with due regard for their diffusion onto the discharge tube walls, and is found as an average over all of the three levels with allowance made for their degeneracy (Fig. 2.10). The term Tn_m^2 allows for those collisions between metastable atoms which result in the ionization of one of them. S_m stands for the electronic ionization coefficient of the quasilevel M, and S' is the coefficient of the electronic excitation of atoms from the quasilevel to the states $2p$. The atoms raised to the $2p$ levels drop back to the quasilevel with a probability of 3/4 to find themselves, with a probability of 1/4, at the $1S_2$ levels and thence drop lower, to the ground state, thus falling out of the ionization process. Table 2.3 lists the values of T, S, and S' for $R = 0.1.$ cm.

The electron-balance equation has the form

$$\frac{dn_e}{dt} = G\left(n_e, n_m, E\right) = \alpha\mu_e E n_e + S_m n_e n_m + \frac{Tn_m^2}{2} - \frac{D^+}{\Lambda^2} n_e = 0. \quad (2.48)$$

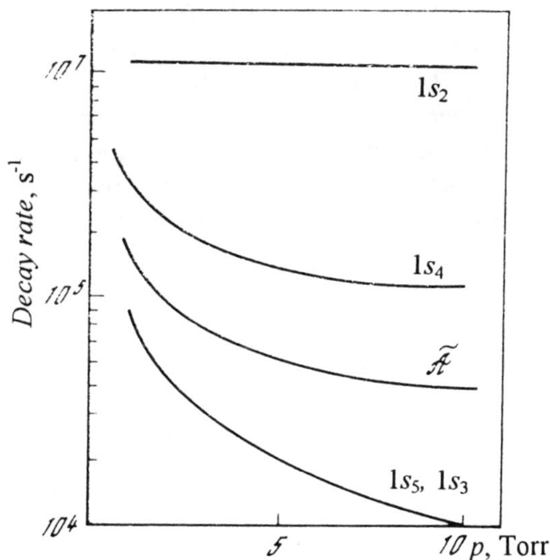

Figure 2.10. Decay rate of the levels 1s as a function of gas pressure [2.13].

The first term of the above equation describes the direct ionization of the gas and is expressed in terms of the Townsend coefficient α and electron drift velocity. Both these quantities strongly depend on the parameter E/p. The second term defines the two-step ionization process proceeding via the quasilevel. The third term describes the collision ionization of metastable atoms, and the fourth defines the ambipolar drift of electrons onto the wall of the cylindrical discharge tube. The values of the quantity $D^{+}(2.4/R)^{2}$ are listed in Table 2.3.

Table 2.3. Values of T, kT_{e}, S_{m}, S', and D^{+}/Λ^{2} at various gas pressure values.

Pressure, Torr	T, 10^{-9} $cm^{3}{\cdot}s^{-1}$	kT_{e}, eV	S_{m}, 10^{-8} $cm^{3}{\cdot}s^{-1}$	S', 10^{-7} $cm^{3}{\cdot}s^{-1}$	D^{+}/Λ^{2}, 10^{6} s^{-1}
1	2.6	4.7	6.7	6.7	9.5
2	2.6	3.9	5.4	5.4	3.9
5	2.6	2.8	2.9	2.9	1.1
10	2.6	2.2	1.6	1.6	0.44

The discharge current is determined from Eq. (2.18). Given the magnitude of the current, the quantities n_{e}, n_{m}, and E can be found by solving the set of Eqs. (2.47), (2.48), and (2.18).

The resonant radiation raises the atoms from the state $1s_{2}$ to the state $2p_{4}$. The excited atoms drop from the state $2p_{4}$ back to the state $1s_{2}$ with a probability of $\beta = 0.44$. As a result, the population of the quasilevel diminishes. This process may be described, by analogy with the foregoing, by the equation

$$\frac{\partial H}{\partial n_{e}}\Delta n_{e} + \frac{\partial G}{\partial n_{m}}\Delta n_{m} + \frac{\partial G}{\partial E}\Delta E = \beta Q, \qquad (2.49)$$

and the perturbed charged-particle balance equation has the form

$$\frac{\partial H}{\partial n_{e}}\Delta n_{e} + \frac{\partial G}{\partial n_{m}}\Delta n_{m} + \frac{\partial G}{\partial E}\Delta E = 0. \qquad (2.50)$$

Based on Eqs. (2.49, (2.50), (2.38), and (2.5), one can obtain the following expression for ΔE:

$$\Delta E = -Q\frac{\partial F}{\partial n_{\mathrm e}}\frac{\partial G}{\partial n_{\mathrm m}}\left\{\frac{\partial F}{\partial n_{\mathrm e}}\left[\frac{\partial H}{\partial n_{\mathrm m}}\frac{\partial G}{\partial E}-\frac{\partial G}{\partial n_{\mathrm m}}\frac{\partial H}{\partial E}\right]\right.$$

$$\left.+\left[\frac{\partial F}{\partial E}+\frac{L}{z}\right]\left[\frac{\partial H}{\partial n_{\mathrm e}}\frac{\partial G}{\partial n_{\mathrm m}}-\frac{\partial G}{\partial n_{\mathrm e}}\frac{\partial H}{\partial n_{\mathrm m}}\right]\right\}^{-1}. \tag{2.51}$$

Using expression (2.41) for the dynamic resistance of the discharge column and also Eqs. (2.47), (2.18), and (2.51), we may express the optogalvanic response Δi as

$$\Delta i = -\frac{L\Delta E}{z} =$$

$$= \beta Q\frac{\partial F}{\partial n_{\mathrm e}}\frac{\partial G}{\partial n_{\mathrm m}}\left\{1+z\left(\frac{dV}{di}\right)^{-1}\left[\frac{\partial H}{\partial n_{\mathrm e}}\frac{\partial G}{\partial n_{\mathrm m}}-\frac{\partial G}{\partial n_{\mathrm e}}\frac{\partial H}{\partial n_{\mathrm m}}\right]\right\}^{-1}. \tag{2.52}$$

And finally using the expressions for F, G, and H, we get the following expression for the optogalvanic current response:

$$\Delta i = \beta Qi\left(S_{\mathrm m}n_{\mathrm e}+Tn_{\mathrm m}\right)\left\{\left[1+z\left(\frac{dV}{di}\right)^{-1}\right]\times\right.$$

$$\left.\left\{\left[S_{\mathrm m}n_{\mathrm e}n_{\mathrm m}+\frac{Tn_{\mathrm m}^2}{2}\right]\widetilde{A}+\left(S_{\mathrm m}-\frac{S'}{4}\right)\frac{n_{\mathrm e}Tn_{\mathrm m}^2}{2}\right\}^{-1}\right\}. \tag{2.53}$$

The experiment was conducted with a normal glow discharge having a positive column length of $L = 13.5$ cm and a radius of $R = 0.1$ cm. The resistance of the ballast resistor was $z = 200$ kΩ. The concentration of the excited atoms was determined by the radiation absorption method and that of electrons, by measuring the discharge current. Figure 2.11 presents the concentrations of the excited atoms and electrons and also the dynamic resistance of the discharge as a function of the discharge current.

The authors of [2.13] determined the magnitude of the optogalvanic effect as the ratio between the change in the power dissipated across the ballast resistor and the absorbed radiation power of frequency ω:

$$\Phi = z2i\Delta i/\hbar\omega Q\pi R^2 L. \tag{2.54}$$

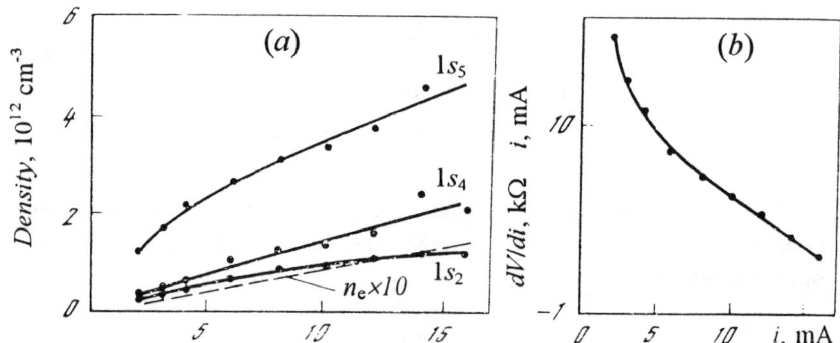

Figure 2.11. (*a*) Space-averaged densities of excited atoms and electrons and (*b*) dynamic resistance of the discharge as a function of the discharge current at a gas pressure of $p = 1$ Torr [2.13].

Figure 2.12 presents theoretical and experimental values of the quantity Φ as a function of the discharge current at various gas pressure values.

The experimental Φ values in [2.13] were obtained by substituting the measured Δi and Q values into Eq. (2.54). To construct semiempirical curves, the authors used in Eq. (2.54) experimental Q values and theoretical Δi values. The theoretical calculation of the Δi values in [1.4] was in turn performed using experimentally measured values of T_e, n_e, populations of the lower-block levels $1s$, and dynamic resistance of the discharge, dV/di.

The theoretical curves of [2.8] were obtained from the closed system of Eqs. (2.5), (2.11), (2.15), (2.16), and (2.18) for the case of normal glow discharge subject to the condition $z = R_e \gg R_s$. In that case, it follows from Eq. (2.5) that $\Delta i = -L\Delta E z^{-1}$ and $\Phi = -[2i/(\hbar\omega Q\pi r^2)]\Delta E$. The quantity Q is found by the Bouguer law: $Q = F[1 - \exp(-k_0 L)]$. The quantity F corresponds to a power of 4 mW. Within the framework of the five-level model [2.13], the magnitude of the optogalvanic signal Δi is proportional to F, provided that the electron temperature remains constant. The absorbed laser power is also proportional to F, and so the quantity Φ is independent of the external radiation power. But in the general case, where a more precise account is taken of the atomic structure and the dependence of electronic processes on T_e [2.8], the quantity Δi and hence Φ depend on the a more precise account is taken of the atomic structure and the dependence of electronic processes on T_e [2.8], the quantity Δi and hence Φ depend on the external radiation power. One can see from Fig. 2.12 that

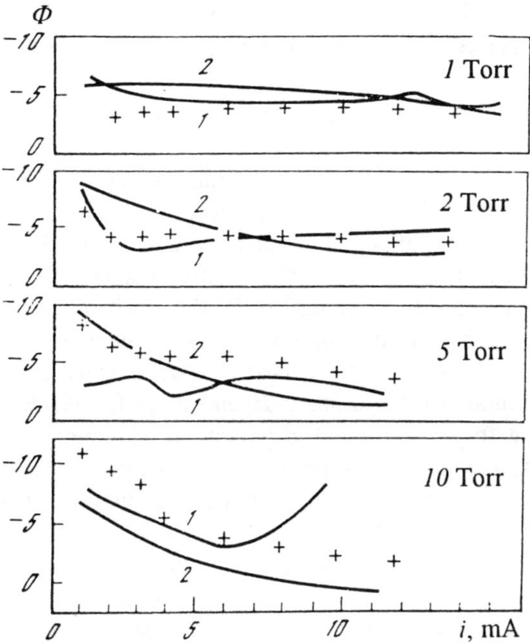

Figure 2.12. Quantity Φ as a function of the discharge current at various gas pressure values for $R = 1$ mm. 1 – theory [2.13]; crosses – experiment [1.4]; 2 – theory [2.8]. Transition $1s_5 \rightarrow 2p_4$, $\lambda = 5944.8$ Å.

the difference between the model data and experimental results becomes manifest at high gas pressures and discharge currents. The authors of [2.13] believe that this is associated with the effect of the bulk recombination of charged particles.

This model was later generalized in [214]. It is a well-established fact that the cathode and anode regions play an important part in glow discharge and are actually responsible for its development. The role of each of these regions was studied in [2.15] by exciting Ar atoms and measuring the optogalvanic signal, the spatial variations of the charge being detected by means of a probe.

2.4 LIGHT-INDUCED CHANGES IN ABNORMAL GLOW DISCHARGE CHARACTERISTICS

In the preceding section, we have considered a stationary model of the optogalvanic effect in a normal glow discharge with its positive column subject to an optical perturbation. Generally speaking, to construct a nonstationary model of the optogalvanic effect in a normal glow discharge is, to our view, not a principal problem, provided the kinetics of the elementary processes occurring in the positive column are known. The point is that the optical perturbation of the positive column is fully reflected by the changes in the current flowing in the external circuit, for, as already stated in Sec. 2.1, the electrical characteristics of the electrode regions in a normal discharge are independent of the discharge current. The situation with an abnormal discharge is different. In that case, introducing perturbations in the positive column causes alterations in the cathode region conditions. And it is exactly this situation that is studied in the present section.

Consider a nonstationary optogalvanic effect in an abnormal glow discharge with its positive column exposed to a resonance radiation. To develop a nonstationary distributed-parameter model of a glow discharge is an extremely difficult problem practically unsolved to date [2.2]. The authors of [1.4] suggested a simple lumped-parameter model described by ordinary differential equations.

When the output resistance of the supply source is low, the voltage drop V across the discharge tube remains constant and is defined by Eq. (2.2).

The current in the discharge circuit is governed by the electrode voltage drops and the emission ability of the cathode, and the positive column plays the part of a wire connecting together the anode and cathode regions. The relation $V_c \gg V_a$ usually holds in discharges. Therefore, the discharge current

$$i \cong f(v_c), \tag{2.55}$$

where f is the volt-ampere characteristic of the discharge.

If the conductivity variation time in the positive column is longer than the current relaxation time in the rest of the gas-discharge circuit, the magnitude of the discharge current adiabatically follows the condition of the positive column, so that one may write

$$i \cong f(V - EL). \tag{2.56}$$

For this condition to be satisfied, it is necessary that the time $\tau_{\parallel}$ it takes for ions to drift through the cathode region should be much shorter than the time τ_a of their ambipolar diffusion to the discharge tube wall:

$$\frac{\tau_{\parallel}}{\tau_a} \approx \left(\frac{l}{\Lambda}\right)^2 \frac{1}{\mu_i} \frac{D^+}{V_c} \ll 1, \tag{2.57}$$

where μ_i is the ion mobility, l is the cathode region length, $\Lambda = R/2.4$, R is the discharge tube radius, and D^+ is the ambipolar diffusion coefficient.

We will also describe the plasma of the positive column on the basis of a lumped-parameter model. The external radiation uniformly excites the resonant gas – an admixture to the buffer gas which governs particle transport and diffusion processes. The population n_2 of the excited resonance state of the atom is defined by expression (2.11). Atomic collisions are not taken into consideration.

If the governing contribution to ionization comes from the admixture, the charged particle balance equation has the form of (2.16) with the diffusion term defined by expression (2.33).

The current i in the positive column region of discharge circuit is defined by expression (2.18).

Substituting the electron concentration found from Eq. (2.18) into Eq. (2.16), we get

$$\frac{1}{i}\frac{di}{dt} - \frac{1}{E}\frac{dE}{dt} = \sum_{m=1,2} S_m n_m - \frac{D^+}{\Lambda^2}. \tag{2.58}$$

The energy balance equation defining the electron temperature T_e has the form of expression (2.15).

By exciting the gas, the resonance radiation changes its ionization rate and hence the electron temperature T_e. The strength of the axial electric field will be affected by two factors: the change of the ionization rate and the increase of the role of superelastic collisions in the energy balance. And since the voltage drop V is constant, this must lead to a change in the voltage drop in the electrode regions and a change in the current flowing in the gas-discharge circuit.

Thus, in the model under consideration, the light-induced change in the conductivity of the positive column brings about a voltage redistribution in the glow discharge and a change in the discharge current.

The system of Eqs. (2.11), (2.15), (2.17), (2.18), 2.56), and (2.58) defines the behavior of n_1, n_2, n_e, i, E, and T_e and models the dynamic optogalvanic phenomenon. By and large, the magnitude of the optogalvanic response depends on the reaction of the gas-discharge circuit to the change in the conductivity of the positive column plasma caused by the action of the resonance radiation.

Concrete calculations for the given system with a small number of energy levels were performed for conditions corresponding to the experiment conducted in [1.4]. The resonance gas was sodium vapor, and the buffer gas, neon. The sodium atoms were excited in a transition with a wavelength of $\lambda = 589$ nm. The laser pulse used for the purpose had a duration of $\tau \cong 10^{-8}$ s and a peak power of some 40 kW. The density of sodium atoms was $n_a \cong 2\times10^{14}$ cm^{-3}, and the neon pressure, 1 Torr. The discharge glowed in a tube 8 mm in diameter, the length of the positive column being $L = 10$ cm. The discharge current equaled 16 mA.

Under these conditions, the laser pulse duration τ is much shorter than the characteristic times of all the processes responsible for the population of the excited state. It follows from Eq. (2.11) that after the passage of the laser pulse the population n_2^* of the excited state will be expressed as

$$n_2^* = [n_a\sigma_{12}/(\sigma_{12} + \sigma_{21})]\{1 - \exp[-F(\sigma_{21} + \sigma_{12})\tau]\} + n_2(0). \qquad (2.59)$$

Here $n_2(0)$ is the concentration of the excited atoms corresponding to the steady-state glow of the discharge prior to the passage of the laser pulse. In calculations, the magnitude of the population n_2^* was taken to serve as the initial value. The initial conditions for all the other variables were found from the stationary solution of the set of equations describing the model of the discharge.

A single laser pulse contained some 5×10^{14} photons. The optical density of the medium in the irradiation direction amounted to around 10^6 at the center of the transition profile for the line $\lambda = 589$ nm. The exciting radiation linewidth was around 0.1 Å. Thus, the medium will absorb radiation detuned from the line center by almost four Doppler profiles some 0.02 Å wide. For this reason, conditions were created in the experiment providing for the saturation of the transition practically throughout the discharge volume containing some 5×10^{14} normal atoms.

It can be seen that in the above experimental conditions the initial assumption as to the "adiabatic" character of the discharge current, (2.57), holds true and $\tau_{||} \leq 0.1\tau_a$.

The calculation for the unperturbed positive column yields the following values: $T_e(0) = 2$ eV, $E(0) = 1.8$ V·cm^{-1}, $n_e = 6 \times 10^{10}$ cm^{-3}. After the passage of the laser pulse, $n_2^* \cong 0.6 n_a$.

To compare between the calculation results and experimental data, use should be made of data on the volt-ampere characteristic of the cathode region of the discharge in accordance with Eq. (2.56). This problem was solved experimentally in [1.4] by means of a specially designed discharge tube with a variable discharge zone length, which made it possible to exclude the positive column zone from the discharge. It turned out that in conditions close to those of the experiment described in [1.4] the volt-ampere characteristic could be approximated well enough by a linear relationship of the form

$$f(V - EL) = b + \chi(V - EL - d), \quad V - EL \geq D. \tag{2.60}$$

At a voltage of $V = 158$ V, the approximation parameters are $d = 110$ V, $b = 10^{-3}$ A, and $\chi = 0.5 \times 10^{-3}$ A/V.

Figures 2.13a and b show the temporal behavior of the quantities n_2, E, n_e, and T_e found theoretically. The theoretical and experimental behavior of the discharge current i is presented in Fig. 2.13c.

Following the instantaneous radiative excitation of the resonant gas, the gas-discharge system falls out of its steady-state condition, and the subsequent behavior of its response is governed by the relaxation of the perturbation, which depends on various elementary processes. The increase of the role of stepwise ionization, caused by the excitation of sodium atoms in the positive column, leads to a reduction of the electric field strength E, an increase of the cathode potential drop, and a rise of the discharge current. At the same time, T_e decreases, and so do the coefficients S_1 and S_2. The current rises till the instant t' such that the right-hand side of Eq. (2.58) is equal to zero:

$$S_1(T_e(t'))n_1(t') + S_2(T_e(t'))n_2(t') = D^+(t)/\Lambda^2. \tag{2.61}$$

At the instant $t = t'$ the magnitude of the current response reaches its maximum, whereas E and T_e are minimal. It can be seen from Eq. (2.16) that the increment of n_e depends on n_2. Therefore, as the extent of excita-

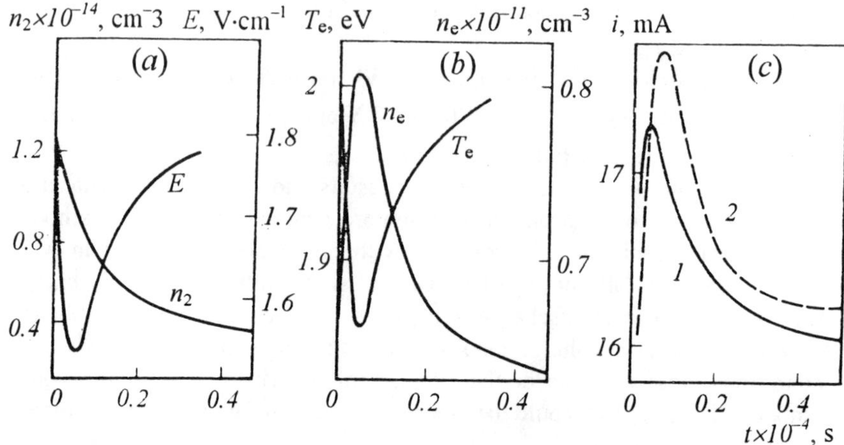

Figure 2.13. Temporal behavior of (*a*) n_2 and E, (*b*) T_e and n_e, and (*c*) i. *1* – theory; *2* – experiment.

tion of the gas grows higher, the magnitude of the current response increases.

Further relaxation of the current is governed by the slowest processes included in the model under consideration. These are spontaneous decay with due regard for radiation trapping, quenching electron impact, and ambipolar diffusion. At low sodium vapor pressures, when the trapping of radiation is insignificant, the excited state decays rapidly, and the reduction of the current response is associated with the ambipolar diffusion of the charged particles. As the concentration n_a of the sodium atoms is increased, the effective radiative decay time of the excited state grows in excess of the ambipolar diffusion time of the charged particles, and the temporal fall of the current will be determined by the reduction of the concentration $n_2(t)$. The quenching electron impact rate becomes commensurable with the radiative decay rate of the excited state at $n_a \geq$ 2×10^{14} cm^{-3}. In the experimental situation under study, the radiative decay and electronic deactivation rates of the excited state and the ambipolar diffusion rate of the charged particles prove to be commensurable, their order of magnitude being around 10^5 s^{-1}.

So, the simple model considered here describes the experiment quite satisfactorily.

While it is practically sufficient to restrict oneself to a three-level scheme to describe the above case of sodium, to do so, for example, in the case of optogalvanic effect in neon (a widespread case, too) requires a more detailed account of the structure of the atomic terms [2.16].

To analyze the optogalvanic effect in a neon plasma, use was mainly made of the model described in Sect. 2.2.

If the relationship between electron mobility and T_e is only defined in terms of the electron velocity $v_e(T_e)$, Eq. (2.18) may then be rearranged in the form

$$n_e = Ai\sqrt{T_e} \,/\, E, \quad A = \left(1.36 R^2 e\mu_e\sqrt{T_e}\right)^{-1}. \tag{2.62}$$

Considering Eqs. (2.60) and (2.62), we rearrange Eq. (2.16) in the form

$$\frac{dE}{dt} = \frac{iE}{i + \chi EL}\left[\frac{1}{T_e}\frac{dT_e}{dt} + \frac{D^+(T_e)}{\Lambda^2} - \sum_{m=1}^{r} S_m(T_e)n_m(n_e, T_e)\right]. \tag{2.63}$$

From Eqs. (2.60) and (2.62) we have

$$\frac{di}{dt} = \frac{\chi ELi}{i + \chi EL}\left[\sum_m S_m(T_e)n_m(n_e, T_e) - \frac{D^+(T_e)}{\Lambda^2} - \frac{1}{T_e}\frac{dT_e}{dt}\right]. \tag{2.64}$$

In the above equation, the first and second terms define the discharge current change caused by the creation and annihilation of electrons, and the third term describes the dependence of the discharge current on the variation of the electron mobility.

The change in the populations of the excited states is defined by Eqs. (2.11), (2.15), (2.17), (2.60), (2.62), and (2.64). The system of equations describing the dynamic optogalvanic effect in a neon plasma was numerically solved by the Geer technique.

The dynamic optogalvanic effect in an abnormal neon discharge was studied in [2.16]. The discharge was implemented at a neon pressure of 1 Torr and discharge currents of 5–20 mA in a 8-mm-diameter tube with a positive column length of $L = 1$ cm. The parameter χ of the volt-ampere characteristic was determined by the method described above. The laser parameters were similar to those described in connection with the experiment with neon [1.4].

The authors of [2.16] experimentally observed three types of optogalvanic response (Figs. 2.14a, b, and c, I) in the neon transitions $1s_i \rightarrow 2p_j$. The same figures also present theoretical data (II).

Let us consider the formation dynamics of the optogalvanic signal. In view of the short duration of the laser pulse, its action boils down to the initial additional excitation of the neon atom to the level $2p_j$, which causes an increase in (1) the temperature of electrons upon their superelastic collisions with the excited atoms in the $2p_j$ state and (2) the number of ionization events on account of the additional population of the level $2p_j$. Calculations show that the effect of the second process on the formation of current in the case of excitation of the transitions $1s_5 \rightarrow 2p_8$ and $1s_4 \rightarrow 2p_8$ proves stronger, i. e., the increase of the atomic ionization events predominates over the variation of the electron mobility and ambipolar diffusion. As a result, the current grows higher to form the beginning of the first positive pulse. In the case of $1s_2 \rightarrow 2p_2$ transition, the rise of the ionization rate proceeds at a slower pace because of the isolated character

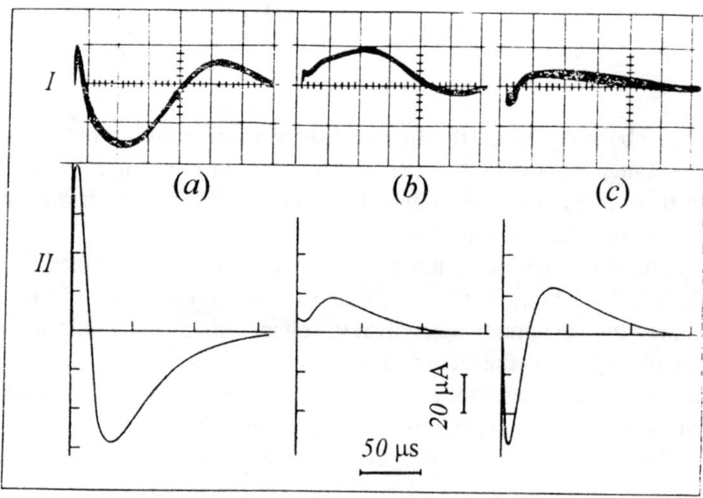

Figure 2.14. Optogalvanic current responses in neon (p_{Ne} = 1 Torr, i (0) = 10 mA). I – experiment; II – theory. Transitions excited: (a) $1s_5 \rightarrow 2p_8$; (b) $1s_4 \rightarrow 2p_8$; (c) $1s_2 \rightarrow 2p_2$.

of the level $1s_2$ (it lies higher than the other $1s_i$ levels, and the coefficient of ionization from this level is the highest in the entire lower block), so that the predominant part is played by the changes in the electron mobility and ambipolar diffusion, the current drops down, and the beginning of the negative response is thus formed. The electron temperature invariably rises on account of superelastic collisions. Figures 2.15 and 2.16 present data on the behavior of the electron temperature and concentration, the electric field, and the population of the atomic levels.

As a result of the spontaneous decay of the $2p_j$ state to the lower-block levels following its laser excitation $1s_i \rightarrow 2p_j$, their total population changes only slightly because the spontaneous decay probability of the $2p_j$ level is substantially higher than its electronic ionization probability. But the populations of the $1s$ levels undergo redistribution relative to their values not disturbed by the radiation. Accordingly, as the higher- or lower-lying levels are populated (relative to the initial $1s_i$ state), one can note a decrease or an increase in the effective ionization potential of the block of levels $1s$. When the external radiation acts upon metastable states ($1s_3$, $1s_5$), the subsequent spontaneous decay of the excited $2p_j$ level causes the populations of the resonance states ($1s_2$, $1s_4$) to grow in excess of their unperturbed values.

Insofar as the lifetime of the resonance levels is shorter than that of their metastable counterparts, the decay of the levels $1s_2$ and $1s_4$ results in the total population of the states in the lower block becoming smaller than the unperturbed populations. If the initial state is a resonance one, the total

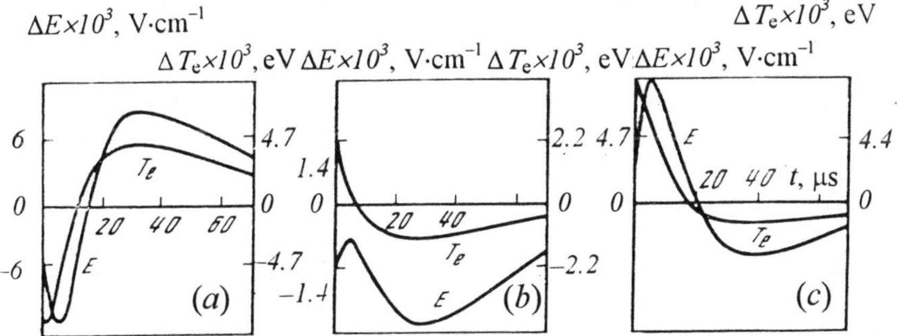

Figure 2.15. Variation of T_e and E in optogalvanic signals. $T_e(0) = 2.58$ eV; $E(0) = 2.93$ V·cm^{-1}; (a) $1s_5 \rightarrow 2p_8$; (b) $1s_4 \rightarrow 2p_8$; (c) $1s_2 \rightarrow 2p_2$.

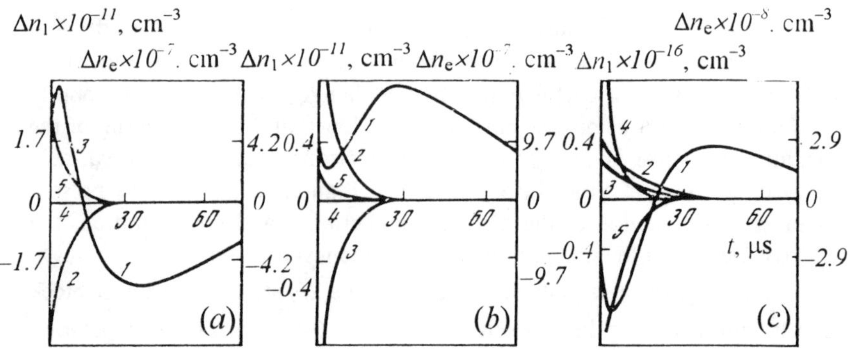

Figure 2.16. Variation of Δn_i and Δn_e in optogalvanic signals. (a) $1s_5 \to 2p_8$; (b) $1s_4 \to 2p_8$; (c) $1s_2 \to 2p_2$. $1 - \Delta n_e$; $2 - \Delta n_2(1s_5)$; $3 - \Delta n_3(1s_4)$; $4 - \Delta n_4(1s_3)$; $5 - \Delta n_5(1s_2)$.

population of the lower-block levels will be greater than the unperturbed populations. Figure 2.16 presents theoretical temporal variation curves for the populations of the resonance and metastable states, the total population of the lower-block levels, and electron concentration.

The behavior of the discharge current following the spontaneous decay of the state $2p_j$ depends on the variation of the populations of the levels $1s$ and the electron temperature, the role of the first factor being dominant. For the transition $1s_5 \to 2p_8$ (Fig. 2.16a), the first to become manifest is the reduction of the ionization potential (since the $2p_j$ level mainly decays to $1s_4$ and $1s_5$), and the discharge current continues to grow. Further, the total population of the levels in the block $1s$ is influenced by the depletion of the resonance states as a result of their spontaneous decay (see Fig. 2.16a), and the current starts decreasing. Finally, the population of the metastable states comes into play, and the current, while increasing, reaches its initial value. In the case illustrated in Fig. 2.16b, the ionization potential increases, the current starts dropping and continues to do so until the total population of the lower-block levels becomes influenced by the resonance state $1s_4$ becoming populated against the background of the slow decay of the metastable level. At that instant, the current starts to grow, and its reaching the initial value is governed by the further relaxation of the population of the metastable states. In Fig. 2.16c, the effective ionization potential of the lower block of levels also increases, the current grows further, and the picture subsequently becomes similar to the preceding one.

Thus, these theoretical models of the optogalvanic effect occurring in the positive column of normal and abnormal glow discharges satisfactorily describe the changes observed in the electrical parameters of the discharge, at least for comparatively simple atomic systems.

Of course, it would be premature to state that the given models can already be taken as a basis for the solution of a wide range of inverse diagnostic problems in optogalvanic spectroscopy. Although the development of methods for diagnosing nonequilibrium gas-discharge plasmas by means of optogalvanic spectra seems realistic in principle (some examples will be presented elsewhere), serious additional investigations still remain to be undertaken here.

2.5 OPTOGALVANIC EFFECT IN HOLLOW-CATHODE AND OBSTRUCTED DISCHARGES

This type of discharge is widely used in studies of atomic optogalvanic spectra. Hollow-cathode lamps designed for the excitation of resonance transitions in most atoms, primarily for atomic absorption spectroscopy and analysis purposes are commercially available.

The hollow-cathode discharge theory is not fully developed, and the interpretation of the optogalvanic effect in this type of discharge is usually based on qualitative or semiempirical considerations. Specifically, the question of which mechanism – selective ionization or the heating of electrons – is the principal contributor to the optogalvanic effect in the hollow-cathode discharge is being discussed.

A phenomenological approach was attempted in [2.17–2.19] to describe this case as being the result of selective ionization. The authors introduce the multiplication coefficient β determining the number of electrons emitted by the cathode as a consequence of a set of processes initiated by a single preceding electron. In the steady-state case, $\beta = 1$. The change in the discharge voltage caused by a radiative perturbation is expressed in terms of the initial difference in population between the levels involved, the relaxation times of these levels, and the derivatives of the coefficient β with respect to voltage and level populations. These quantities are specified in order to attain the best fit with the experimental data on the evolution of the optogalvanic effect over a time interval of up to 10^{-4} s. The number of relaxation times to be taken into consideration is substantiated qualitatively. In a number of cases, this approach enables one to describe the evolution of the optogalvanic effect in sufficient detail. Its disadvantage is that a great number of relaxation times (up to 4 in the case of neon) cannot be explicitly associated with concrete physical processes, so that the approximation of the experimental relationships is formal to some extent. Additional research is required for the description of the sign-variable optogalvanic effect by means of inverted states. For example, to interpret the response curves presented in Fig. 2.14 for an abnormal glow discharge requires no level population inversion notions to be invoked.

The optogalvanic effect in a hollow cathode was described as the result of the heating of electrons upon collisions of the second kind in, for example, [2.20–2.22]. The estimates presented in [2.22] for the excitation of uranium show that the change of conductivity is mainly associated with the heating of electrons, the relative electron heating $\Delta T_e/T_e$ amounting to some 3×10^{-3}. According to the same estimates, the role of selective ionization is 10–100 times weaker. In favor of the electron heating mechanism are also the experiments conducted in [2.20] on the excitation of U II leading to the simultaneous intensification of the U I emission lines. The mass-spectrometric experiments performed in [2.23] with a hollow cathode in neon, which show the existence of a correlation between the radiation-induced changes in the discharge voltage and the density of Ne^+, argue against the exclusive influence of the heating of electrons upon the formation of the optogalvanic effect. The authors of [2.24] object against the inference made in [2.25] that the efficiency of the optogalvanic effect is independent of the ionization potential of the atom, an argument in favor of the electron heating mechanism.

The authors of [2.26] studied the optogalvanic effect in a hollow-cathode lamp with the cathode surface exposed to high-power laser pulses (~ 1 MW, 10 ns). The effect had a nonresonance character; its magnitude depended on the material of the cathode and grew rapidly with increasing radiation power. These authors interpreted the experimental results observed as a multiple-quantum effect.

There are many observations on the influence of the discharge conditions on the magnitude of the optogalvanic effect. An important point is the change of the polarity of the optogalvanic signal consequent upon alteration of the discharge conditions even for a fixed transition.

Figure 2.17 shows the relationship between the optogalvanic signal (ΔV) and the discharge current (a hollow cathode in neon) [2.27] for three radiation wavelengths. To illustrate, the optogalvanic effect caused by a radiation with a wavelength of $\lambda = 603$ nm changes sign two times. The authors of [2.27] put forward a qualitative explanation of the above relationships, based on a generally five-level scheme, which is illustrated in Fig. 2.18. In this figure, the symbol M denotes a metastable level, H

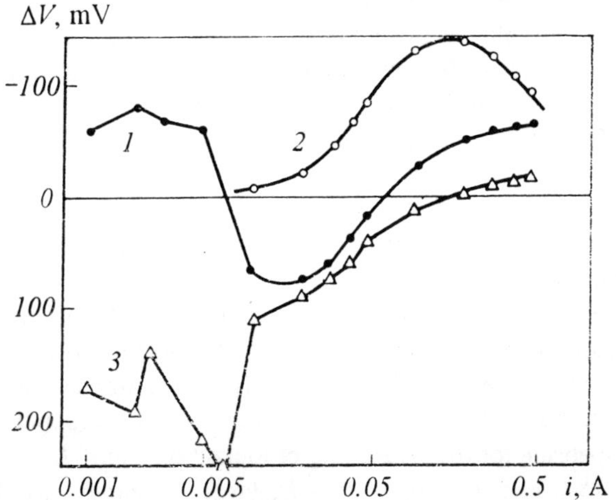

Figure 2.17. Optogalvanic effect ΔV as a function of the discharge current in a hollow-cathode discharge in Ne for three radiation wavelengths [2.27]. $1 - \lambda = 603$ nm; $2 - \lambda = 576.4$ nm; $3 - \lambda = 588.2$ nm.

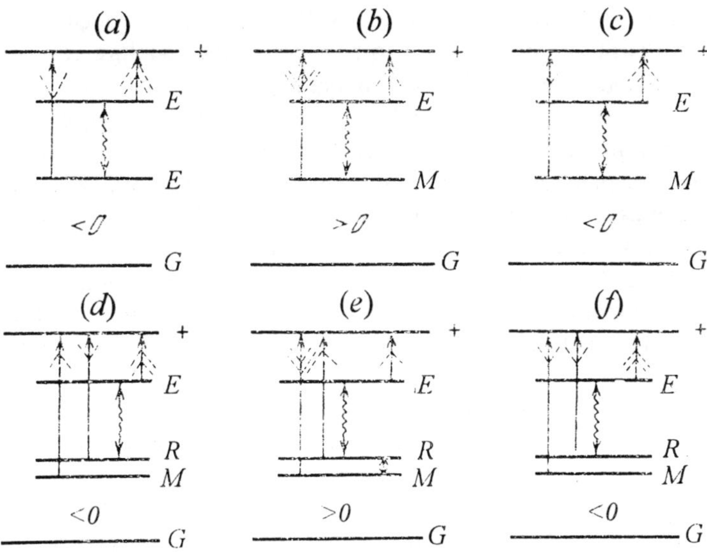

Figure 2.18. Illustrating the sign reversal of the optogalvanic effect consequent upon alteration of the discharge current [2.27].

stands for a highly excited level, R is a resonance level, and G is the ground state. The plus sign indicates the ionization limit. The solid straight arrows indicate the ionization and deionization processes taking place in the absence of the external radiation, the wavy ones, the transition under the effect of the radiation, and the dashed arrowheads show the direction in which the rates of the processes vary as a result of irradiation, the double ones indicating faster processes.

Figure 2.18a. The transition $H \rightarrow H$. Ionization proceeds faster from the higher-lying level, $\Delta V < 0$.

Figure 2.18b. The transition $M \rightarrow H$, small discharge current. The external radiation raises the atom from the metastable to the highly excited state. The decay of the level H can take its course by various channels, including those passing the level M. The increase of the population of H does not compensate for the weakening of ionization from M, $\Delta V > 0$.

Figure 2.18c. The transition $M \rightarrow H$, large discharge current. As the discharge current is increased, the population of the level M gets saturated, but the population of the level H continues to grow, so that the rate of ionization from H is higher than that from M. By and large, the decay of the level M and the population of the level H under the effect of the radiation increase the ionization rate, $\Delta V < 0$.

Figure 2.18d. The transition $R \rightarrow H$, small discharge current. The electron density is low and fails to provide for the coupling between the levels R and M. The case is similar to that of Fig. 2.18a, $\Delta V < 0$.

Figure 2.18e. The transition $R \rightarrow H$. The electron density is sufficient for the mixing of the levels R and H, but is still too low to provide for a noticeable population of H. As in the case of Fig. 2.18b, the weakening of ionization as a result of the decay of the level R (and the level M as well) is not compensated for by the increase of ionization from the level H, $\Delta V > 0$.

Figure 2.18f. The transition $R \rightarrow H$. The electron density is high enough to provide for both the mixing of the levels R and M and a high population of the level H. The situation is similar that of Fig. 2.18c, that is, $\Delta V < 0$.

The above diagrams explain the relationships presented in Fig. 2.17. The group of levels $1s$ in neon consists of two metastable and two resonance levels (see Fig. 2.6). The separation of the nearest pairs among the three lowermost levels amounts to approximately 0.05 eV, which makes for their mixing by electrons. The transition at a wavelength of $\lambda = 576.4$ nm always leads to $\Delta V < 0$ (Fig. 2.18a). The transition at $\lambda = 588.2$ nm corresponds to the case of Fig. 2.18b at small discharge currents and to that of Fig. 2.18c at large currents. As the discharge current is increased, the transition at $\lambda = 603$ nm results in the optogalvanic effect behaving as in Figs. 2.18*d* through *f.* The variation of the electron density in the discharge explains the dependence of the optogalvanic effect on the position of the irradiating beam relative to the discharge axis. Examples of such relationships are presented in Fig. 2.19 for the 588.2-nm line of Ne.

Summarizing, it is evident that further research is necessary in order to obtain a clearer understanding of the optogalvanic effect in hollow cathode discharges.

Experiments with obstructed glow discharges are also of interest. The length of such a discharge is short, and its overwhelming portion is constituted by the cathode drop region [2.28]. Ionization is hindered, and current is mainly carried by ions which are produced by electrons. For this reason, even an insignificant change in the rate at which electrons are being expelled from the cathode noticeably affects the discharge current. The effect of irradiation on the current may manifest itself in the annihilation of metastable atoms active in the expulsion of electrons from the cathode surface. This circumstance was used by the authors of [2.29].

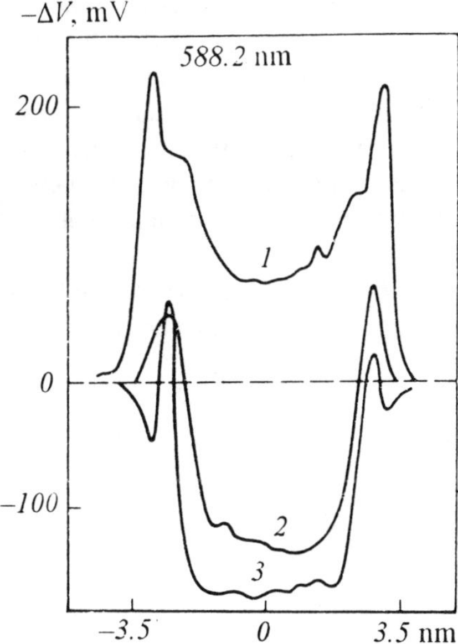

Figure 2.19. Optogalvanic effect ΔV as a function of the position of the irradiating beam relative to the hollow cathode axis. Neon pressure 5 Torr [2.27]; discharge current: *1* – 0.05 A; *2* – 0.2 A; *3* – 0.4 A.

They studied the optogalvanic effect for the 594.5-nm line of neon. The discharge gap was 1 cm wide, and the electrodes were in the form of plates 3 by 10 cm^2 in area. Large plates were necessary in order that the weak absorption of the radiation illuminating the discharge in a direction normal to the flow of the discharge current could be detected. The quantity Φ (the ratio between the optogalvanic signal power dissipated in the external circuit and the absorbed radiation power) was found to be dependent on the electrode material and much higher than in the positive column. Of the electrode materials studied (steel, stainless steel, copper, brass, molybdenum, aluminum), the greatest effect was achieved with aluminum and brass. While the magnitude of Φ for the 594.5-nm line of neon was between -3 and -6 in the positive column [2.13] and around -10 in the hollow cathode [2.30], it proved to be as high as -400 in the above obstructed discharge. The minus sign bespeaks the reduction of the discharge current upon irradiation.

A sharp enhancement of the optogalvanic effect was also noted in [2.31] when recording the spectra of the molecular ions N_2^+ and CO^+ in the cathode dark space region of a discharge. The sensitivity of the optogalvanic technique in this region far exceeded that of the laser-induced fluorescence method, but was inferior to it in the negative glow region (ion concentration around 5×10^9 cm^{-3}). The authors associate the high optogalvanic ion detection sensitivity with the effect of excitation on the mobility of the ions, the ion-neutral molecule charge-exchange cross section of the excited ions being smaller than that of their unexcited counterparts.

It was shown in [2.32] that in the cathode regime for low optogalvanic response values, the current variation ΔI may be written in the form

$$\Delta I = \frac{\partial I}{\partial n_d} \Delta n_d + \frac{\partial I}{\partial n_g} \Delta n_g + \frac{\partial I}{\partial n_m} \Delta n_m.$$

The current variation is due to changes in the number of photons emitted from the Crookes dark space and negative glow region, as well as in the density of metastable atoms, n_m.

2.6 OPTOGALVANIC EFFECT IN MOLECULAR GAS PLASMAS

Research into the optogalvanic effect in molecular transitions is being conducted in both the visible and IR regions of the spectrum. The measurement techniques used in the case of visible-region transitions between electronic states are similar to those employed in the studies of atomic spectra. There is also an analogy between the mechanisms responsible for the occurrence of the optogalvanic effect in molecular electronic transitions and atomic transitions.

Continuously broadband-tunable diode lasers and discretely line-tunable molecular gas lasers are used in the IR region. Figure 2.20 shows some exemplary visible and IR optogalvanic molecular spectra. Such spectra can reliably be recorded even with IR diode lasers having so low an output power as around 1 mW.

The vibrational quantum energy of a molecule (~ 0.1–0.3 eV) is much lower than the ionization energy, and the optogalvanic effect mechanism associated with the lowering of the ionization limit in the case of atomic and molecular excitation in the visible region can hardly be effective here. With this in mind, the authors of [1.19, 2.33] suggested a mechanism associated with vibrational relaxation. While suffering decay upon col-

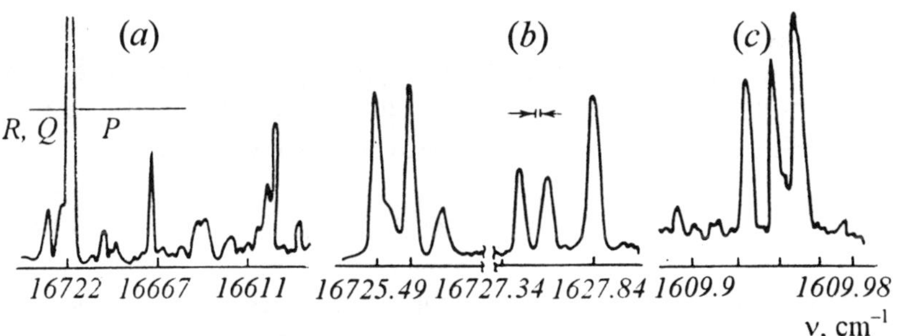

Figure 2.20. Exemplary optogalvanic spectra of molecules in a glow discharge [1.18]. (a) and (b) NH_2 spectra, discharge in NH_3 at a pressure of 1.8 Torr, dye laser; (c) NO_2 spectrum, discharge in NH_2–He mixture at a pressure of 1.2 Torr, diode laser.

lisions, the vibrational-rotational molecular states excited by the external radiation give up their energy to heat, so that the gas density decreases. This cuts down the energy lost by electrons upon electron-molecule collisions, i. e., both an acoustic and an ionization mechanism exist. The persistence of the effect is governed by the relaxation times of the vibrational levels. In [2.33] the phase-frequency characteristics of the optogalvanic effect occurring in the active-medium glow discharge of a CO_2 laser as a result of variation of its intensity modulation frequency were studied. A correspondence was found between the modulation frequencies characteristic of the change in the amplitude and phase of the optogalvanic signal and the relaxation times of the asymmetric and stretching modes of the CO_2 molecule involved in the lasing transition.

Moffat and Smith [2.34] proposed a quantitative fluctuation model for the steady-state optogalvanic effect, based on the above mechanism for the active medium of the CO_2 laser. Subject to analysis was the effect of lasing on the gas temperature T_g and density n_a.

The power imparted to electrons by the external field of strength E is $P = Ej = Eev_d n_e$, where j is the current density and v_d and n_e are the electron drift velocity and density, respectively. In conditions typical of the active media of molecular gas lasers, this power is expended in collisions between the electrons and heavy atoms and molecules. If ε_c is the average electron energy lost in each collision and v_c is the collision frequency, then

$$P = n_e v_c \varepsilon_c. \tag{2.65}$$

Since $v_c \sim n_a$, fluctuations of the density n_a will result in fluctuations of the power imparted by the field to atoms and molecules, provided that the magnitude of ε_c and the loss channels are conserved in collisions:

$$\Delta P/P = \Delta n_a/n_a. \tag{2.66}$$

If the gas volume in the light beam is much smaller than the total volume, the fluctuations of the pressure p can be neglected, and considering that $p = n_a k T_g$, we will have

$$\Delta n_a/n_a = -\Delta T_g/T_g. \tag{2.67}$$

The onset of lasing reduces the energy in the vibrational modes of the molecules and leads to the reduction of the gas temperature as a result of the vibrational-translational energy exchange. To relate quantitatively the changes ΔT_g, Δn_a, and ΔP to the lasing power, it is first necessary to establish the relationship between the electric power deposited in the discharge and the gas temperature. This relationship can be found from the heat conduction equation (for a lengthwise-homogeneous tube with a length of L and a radius of $R \ll L$):

$$\frac{1}{r}\frac{d}{dr}\left(k_{\mathrm{T}}r\frac{dT_{\mathrm{g}}}{dr}\right) + F(r) = 0, \tag{2.68}$$

where $F(r)$ is the radial distribution function of the heat sources.

Within relatively narrow temperature intervals, the thermal conductivity coefficient k_{T} may be approximated by the expression

$$k_{\mathrm{T}}(T_{\mathrm{g}}) = C_1 T_{\mathrm{a}}{}^{\alpha}. \tag{2.69}$$

If the distribution of the heat sources corresponds to a Bessel distribution of the electron density in diffusive discharge glow conditions [2.26], i. e.,

$$F(r) = C_2 J_0(2.4r/R), \tag{2.70}$$

then integrating (2.69) subject to the boundary conditions

$$T_{\mathrm{g}(r=R)} = R, \quad \int_0^R jE\left(1-\eta\right)2\pi r\,dr = P\,/\,L$$

yields [2.35]

$$T_{\mathrm{g}}(r) = T_{\mathrm{g}}(R)\left[1 + \frac{(\alpha+1)(1-\eta)\overline{jE}}{8T_{\mathrm{g}}(R)k_{\mathrm{T}}\left(T_{\mathrm{g}}(R)\right)L}J_0\left(2.4\frac{r}{R}\right)\right]^{1/(\alpha+1)}, \tag{2.71}$$

where $(1-\eta)$ is the proportion of the energy deposited in the discharge that is converted to heat and $\overline{jE}$ is the average taken over the cross section of the discharge tube.

The authors of [2.34] take expression (2.71) in its simplest form for the tube axis, $T_{\mathrm{g}} = T_{\mathrm{g}}(r=0)$, and at $\alpha = 0$, which is confirmed by thermocouple measurements:

$$T_\text{g} = T_\text{g}(R) + T_\text{d} = T_\text{g}(R) + mP/L, \tag{2.72}$$

where T_d is the excess of the gas temperature at the discharge axis over the tube wall temperature, and the coefficient m depends on the gas pressure and composition and is selected empirically. If lasing withdraws a power of Q from the active laser medium discharge, then, according to Eq. (2.72), the relative temperature change will be

$$\Delta T_\text{g}/T_\text{g} = -mQ/[LT_\text{g}(R) + mP]. \tag{2.73}$$

Using the round-trip loss γ in the laser cavity, which includes the output coupling loss, mirror losses, and diffractional loss at the cavity aperture A, the quantity Q may be related to the radiation power density I in the cavity by the following relation (assuming that the beam is radially homogeneous):

$$Q = I(\gamma/2)A. \tag{2.74}$$

In that case,

$$\Delta P/P = \Delta n_\text{a}/n_\text{a} = -\Delta T_\text{g}/T_\text{g} = A\gamma I/2[LT_\text{g}(R)/m + P]. \tag{2.75}$$

Introducing the notion of the optogalvanic effect ratio given by

$$x = P/2[LT_\text{g}(R)/m + P] = 1/2[T_\text{g}(R)/T_\text{d} + 1], \tag{2.76}$$

we may write the following expression for the power dissipated in the discharge under the effect of lasing:

$$P = Ax\gamma I. \tag{2.77}$$

To facilitate comparison between theoretical and experimental data, the quantity I should be related to the lasing power P_L outside the cavity by the well-known relation

$$P_\text{L} = AI(1 - R_\text{m})/(1 + R_\text{m}), \tag{2.78}$$

where R_m is the reflectivity of the output coupling mirror, the rear cavity mirror being totally reflecting. In view of the foregoing,

$$\Delta P = x\gamma P_{\mathrm{L}}(1 + R_{\mathrm{m}})/(1 - R_{\mathrm{m}}), \qquad (2.79)$$

or

$$Q = 1/2\gamma P_{\mathrm{L}}(1 + R_{\mathrm{m}})/(1 - R_{\mathrm{m}}) = \Delta P/2x. \qquad (2.80)$$

If the optogalvanic effect is studied in an amplifying medium with a gain of κ and not in the active medium of a laser, then with the beam power at the amplifier input being $P_{0\mathrm{L}}$, the output power will be

$$P = P_{0\mathrm{g}}e^{\kappa L}. \qquad (2.81)$$

At $\kappa L \ll 1$,

$$P \cong P_{0\mathrm{L}}(1 + \kappa L), \qquad (2.82)$$

and so

$$Q = \Delta P/2x = \kappa L P_{0\mathrm{L}}. \qquad (2.83)$$

The discharge power variation is measured. To illustrate, with the discharge current i being fixed, $\Delta P = i\Delta V$, where ΔV is the discharge voltage variation. The quantities T_{d}, P_{L}, $P_{0\mathrm{L}}$, and P are also measured, and $T_{\mathrm{g}}(R)$ is known. The loss γ is estimated from the known mirror reflectivity R_{m} and mirror and apertural losses, and also from experimentally measured active medium gain and lasing threshold. Figure 2.21a shows the optogalvanic effect ratio x as a function of the power deposition per unit length of the discharge. The solid curve corresponds to x values at the measured T_{d} values, and the dots represent data points calculated by Eq. (2.76) on the basis of experimentally measured ΔP, P_{L}, γ values. Presented in Fig. 2.21b is the power variation ΔP as a function of the power Q withdrawn by the laser radiation from the discharge. The region $Q \geq 1$ W corresponds to the measurements taken in a CO_2 laser and the region $Q \leq 1$ W, to those taken in a CO_2 laser amplifier with discharge conditions similar to those in the laser. The dots represent experimental data points, and the solid curve corresponds to the relationship following from the model. Varying the power by as much as four orders of magnitude does not disturb the agreement between the model and experiment. Thus, its certain apriority and simplification notwithstanding, the thermal perturbation

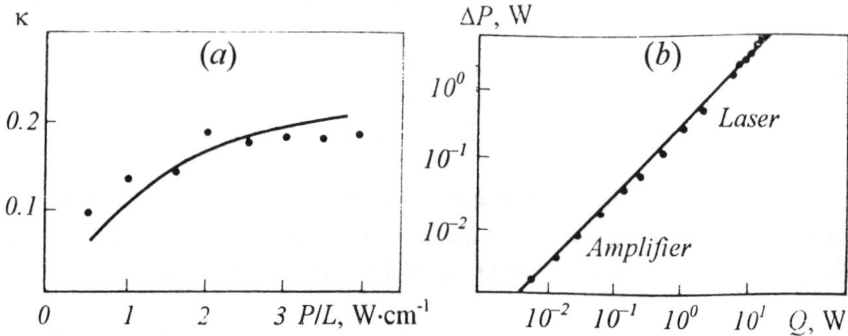

Figure 2.21. Comparison between measurement results and the results predicted by the thermal perturbation model for the active media of a CO_2 laser and amplifier [2.34]. The dots represent experimental data points and the curves, model results. (a) Optogalvanic effect ratio x as a function of the power deposition per unit length of the discharge; (b) variation of the power deposited in the discharge, ΔP, as a function of the power Q withdrawn from the discharge by the laser radiation.

model [1.19, 2.33, 2.34–2.36] is capable of describing the optogalvanic effect in so complex an object as the CO_2 laser active medium, and relation (2.76) may prove useful in determining the parameters characterizing the operation of such a laser. It has been assumed [2.34] that the model is applicable to other molecular IR lasers as well.

The existence of an acoustic mechanism responsible for the occurrence of the optogalvanic effect in IR molecular transitions was proved experimentally in [1.20, 2.37, 2.38] where the effect was investigated while exciting the gas with a resonance radiation outside the discharge region.

Nowicki and Pienkowski [2.39] have examined another model of the optogalvanic effect in the CO_2 laser, which is similar to the models for atomic gas discharges. It presumes that lasing affects the rate of stepwise ionization proceeding via the state $N_2(B^3\Pi)$. This model, however, fails to explain a number of experimental factors and provokes objection [2.40].

Another important representative of high-power IR lasers is the CO laser. It differs from its CO_2 counterpart in that its lasing transitions occur between highly excited vibrational states (usually $v \geq 7$–8), the system itself being an essentially multilevel type.

The steady-state optogalvanic effect in the CO laser was investigated in [2.41, 2.42]. The authors of [2.41] arrived at the conclusion that the mechanism governing the occurrence of the optogalvanic effect in this laser was also associated with thermal perturbations.

The temporal and spectral characteristics of the optogalvanic effect in the CO laser were studied in [2.43]. Subject to investigation was the active element of a CO laser with a CO-N_2-He-Xe mixture at a pressure of some 20 Torr. One of the laser cavity reflectors was either a partially transparent mirror or a diffraction grating. The laser was adjusted to the necessary vibrational-rotational lines by rotating the grating. The voltage of the high-voltage discharge supply source was kept constant, and the optogalvanic effect was observed by monitoring the quantity ΔV – the variable voltage drop across the instrument resistor.

Figure 2.22 presents a typical relationship between the magnitude ΔV of the optogalvanic effect for a number of transitions in the frequency range 1728–1785 cm^{-1} and the lasing power P_L at a frequency of Q-switching equal to 120 Hz. The lasing power was varied by impairing the adjustment of one of the cavity reflectors. It can be seen from the figure that within the measurement error (15–20%) the magnitude of the optogalvanic effect is independent of the transition selected, but is only governed by the lasing power. This result agrees qualitatively with the thermal perturbation model.

The optogalvanic effect in the CO laser was also studied at various frequencies of Q-switching. Some measurement results for the amplitude of the optogalvanic effect are presented in Fig. 2.23 for various lasing lines. The data are joined at the point corresponding to a Q-switching frequency of $v = 120$ Hz. Comparison of the data shows that within the measurement error the relationships for the different lines behave similarly. As the

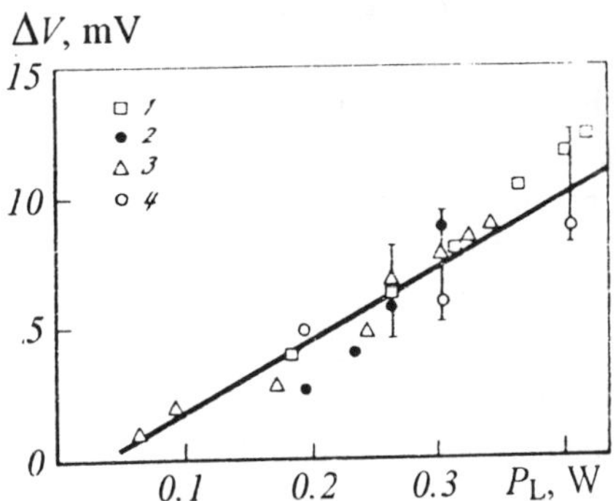

Figure 2.22. Relationship between ΔV and P_L. Lasing frequencies (cm^{-1}): _1_ – 1728; _2_ – 1757; _3_ – 1760; _4_ – 1785.

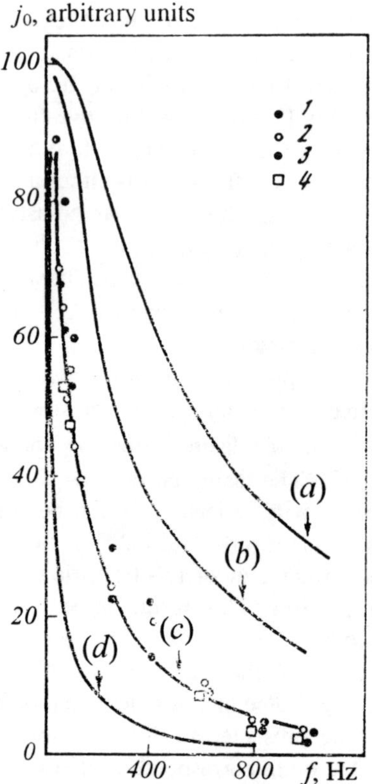

Figure 2.23. Relationship $j_0(\nu)$ in arbitrary units for various characteristic decay times τ_d (s): (a) $\tau_d = 5\times10^{-4}$; (b) $\tau_d = 10^{-3}$; (c) $\tau_d = 3\times10^{-3}$; (d) $\tau_d = 10^{-2}$. Lasing frequency (cm^{-1}): 1 – 1730; 2 – 1740; 3 – 1746; 4 – 1757.

frequency of Q-switching is increased, the signal amplitude lowers to drop by half at $\nu \approx 250$ Hz.

To describe the resultant relationships graphically, a simple model similar to the one employed in [2.44, 2.45] for CO_2 lasers can be used. In this model, the flow of charges in the electric circuit is assumed to obey the equation of an oscillator decaying within a characteristic time of τ_d under the effect of a driving force acting with a frequency equal to the frequency ω of harmonic Q-switching and an amplitude of A_f. Considering that $dq/dt = j$, where j is the discharge current, we have

$$dj/dt + j/\tau = A_f \exp(i\omega t), \tag{2.84}$$

where $\omega = 2\pi v$ and the current amplitude j_0 and phase φ have the form

$$j_0 = A_f \tau / \sqrt{1 + \omega^2 \tau_d^2}, \quad \varphi = \arctan(\omega \tau_d). \qquad (2.85)$$

The relationships $j_0(v)$ are presented in Fig. 2.23 in arbitrary units for some τ_d values. At $v\tau_d \gg 1$, the function $j_0(v) \sim v^{-1}$. Shown in the same figure are also the experimental relationships. As can be seen, the best fit between the experimental relationship and the theoretical curve calculated by Eq. (2.85) is observed at $\tau_d \approx 3\times10^{-3}$ s.

Note that so complex a system as gas discharge features intricately branching connections among its parameters. Exerting some effect on one of the parameters influences the rest of them as well. Acoustic effects consequent upon irradiation are present in atomic gas discharges, too [2.46]. But these effects in atomic gases are to a first approximation concomitant, whereas in molecular gases they constitute a substantial portion of the mechanism responsible for the occurrence of the optogalvanic effect.

The optogalvanic and optoacoustic effects were studied simultaneously in [2.46, 2.47]. To take optoacoustic measurements, a microphone was placed in the discharge gap to detect gas density variations associated with changes in the gas temperature. When irradiating a discharge in CO_2, the temporal behavior of the optogalvanic signal corresponded to that of the optoacoustic signal. But when irradiating a discharge in NH_3 (with an N_2O laser), the shapes of the optogalvanic and optoacoustic signals proved to be different. Whereas the shape of the optoacoustic signal was independent of the zone subject to irradiation, the optogalvanic signal could change both its shape and phase. The author of [2.47] associated this discrepancy with an additional mechanism affecting the formation of the optogalvanic signal in the NH_3 discharge. It might well be that the vibrational excitation of NH_3 molecules by the external radiation raised the probability of electron attachment to them, which naturally affected the conductivity of the discharge plasma.

REFERENCES

2.1. R. C. Webster, *Rev. Sci. Inst.* **54**: 1454–1457 (1983).

2.2. V. L. Granovsky, *Elektrichesky tok v gaze* (Electric current in gases) (Moscow: Nauka, 1971) (in Russian).

2.3. L. M. Biberman, V. S. Vorobyev, I. T. Yakubov, *Kinetika neravnove-snoi nizkotemperaturnoi plazmy* (Nonequilibrium low-temperature plasma kinetics) (Moscow: Nauka, 1982) (in Russian).

2.4. N. Ya. Shaparev, *Rezonansnoye lazernoye upravleniye characteristi-kami gaza i nizkotemperaturnoi plazmy* (Resonance laser control of gas and low-temperature plasma characteristics) (Krasnoyarsk: Candidate's Thesis, 1984) (in Russian).

2.5. J. E. Lawler, *Phys. Rev. A.* **22**: 1025–1033 (1980).

2.6. D. M. Pepper, *IEEE J. Quant. Electron.* **14**: 971–977 (1978).

2.7. M. Maeda, Y. Nomiyama, Y. Miyazoe, *Opt. Comm.* **39**: 64 – 71 (1981).

2.8. V. N. Ochkin, V. A. Pushkaryov, N. Ya. Shaparev, *Modelirovaniye optogalvanicheskikh yavleniy v plazme neona* (Modeling of optogalvanic phenomena in a neon plasma) (Moscow: Preprint FIAN No. 247, 1988) (in Russian).

2.9. V. Martisovits, I. Kosinar, P. Tarabek, S. Veis, in: *Proc. Europhys. St. Conf. on At. and Molec. Phys. Ioniz. Gases.* (Bratislava, 1976): 94–95.

2.10. L. W. Steenhuijsen, *Beitr. aus der Plasmaphysik* **21**: 301–328 (1981).

2.11. Ya. B. Zeldovich, Yu. P. Raizer, *Fizika udarnykh voln i vysokotem-peraturnykh gidrodinamicheskikh yavleniy* (Physics of shock waves and high-temperature hydrodynamic phenomena) (Moscow: Nauka, 1966) (in Russian).

2.12. R. J. Bickerton, A. Von Engel, *Proc. Phys. Soc. London. B.* **9**: 468–481 (1956).

2.13. D. K. Doughty, J. E. Lawler, *Phys. Rev. A.* **28**: 773–780 (1983).

2.14. A. Sasso, M. Ciossa, E. Arimondo, *J. Opt. Soc. Am.* **B5**: 1481–1489 (1988).

2.15. Z. B. Zhu, E. H. Piepmeir, *Spectrochim Acta Pt. B – At. Spect.* **49**, No. 12–14: 1775–1785 (1994).

2.16. N. Ya. Shaparev, N. K. Zaitsev, V. A. Pushkaryov, *Dinamicheskiy optogalvanicheskiy effekt v plazme neona* (Dynamic optogalvanic effect in a neon plasma) (Krasnoyarsk: L. V. Kirenskiy Physics Institute, Preprint No. 274, 1984) (in Russian).

2.17. G. Frez, S. Lavit, E. Miron, *IEEE Quant. Electr.* **15**: 1328–1332 (1979).

2.18. R. Shuker, A. Ben-Amar, G. Erez, *Opt. Comm.* **42**: 29–33 (1982).

2.19. A. Ben-Amar, G. Erez, R. Shuker, *J. Appl. Phys.* **54**: 3688–3698 (1983).

2.20. C. Dreze, Y. Demers, J. M. Gagne, *J. Opt. Soc. Amer.* **72**: 912–917 (1982).

2.21. J. M. Gagne, Y. Demers, P. Pianarosa, C. Dreze, *J. de Phys.* **44**: 355–366 (1983).

2.22. R. A. Keller, B. E. Warner, E. F. Zalewski *et al.*, *J. de Phys.* **44**: 23–33 (1983).

2.23. K. C. Smith, R. A. Keller, E. F. Crim, *Chem Phys. Lett.* **55**: 473–477 (1978).

2.24. N. S. Kopeika, *Appl. Opt.* **21**: 3839–3991 (1982).

2.25. R. A. Keler, E. F. Zalewski, *Appl. Opt.* **19**: 3301–3305 (1980).

2.26. Yin Lifeng, Hu Qiguan, Su Haizheng, Lin Fucheng, *Chinese Physics* **5**: 973–979 (1985).

2.27. E. M. van Veldhuisen (Eindhovev: Thesis, 1983).

2.28. A. Engel, *Ionizovannye gazy* (Ionized gases) (Moscow: GIFML, 1959).

2.29. D. K. Doughty, J. E. Lawler, *Appl. Phys. Lett.* **42**: 234–236 (1983).

2.30. E. F. Zalewski, R. A. Keller, *J. Chem. Phys.* **70**: 1015–1026 (1979).

2.31. R. Walkup, R. W. Dreufus, Ph. Avouris, *Phys. Rev. Lett.* **50**: 1846–1849 (1983).

2.32. De Marinis, E. A. Sasso, E. Arimondo, *J. Appl. Phys.* **63**: 649–654 (1988).

2.33. S. Moffat, A. L. S. Smith, *Opt. Comm.* **37**: 119–123 (1981).

2.34. S. Moffat, A. L. S. Smith, *J. Phys.* **D17**: 59–70 (1984).

2.35. V. N. Ochkin, S. Yu. Savinov, Yu. B. Udalov, *O kharaktere usredneniya pri izmereniyakh temperatury gaza po elektronno-kolebatelno-vrashchatelnym spektram v razryadnykh trubkakh* (On the character of averaging in measuring gas temperatures in discharge tubes by means of electronic-vibrational-rotational spectra) (Moscow: Preprint FIAN No. 36, 1977) (in Russian).

2.36. D. C. A. Dutu, V. Draganescu, N. Comaniciu, D. C. Dmitrius, *Rev. Roum. Phys.* **30**: 127–130 (1985).

2.37. R. E. Munchausen, R. D. May, G. W. Hill, *Opt. Comm.* **48**: 317–321 (1984).

2.38. N. K. Zaitsev, N. Ya. Shaparev, *JhTF* **55**: 218–220 (1985).

2.39. R. Nowicki, J. Pienkowski, *J. Phys. D.* **75**: 1165–1180 (1982).

2.40. A. L. S. Smith, S. Moffat, *J. Phys. D.* **17**: 71–78 (1984).

2.41. M. J. Kavaya, R. T. Menzies, U. P. Oppenheim, *IEEE Quant. Electr.* **QE-18**: 19–21 (1982).

2.42. Wang Yumin, Gui Zenxin, Zhang Shunyi, *Chin. Phys.* **2**: 799–803 (1982).

2.43. S. V. Gorbovsky, V. N. Ochkin, A. G. Sviridov, *Optogalvanicheskiy effekt v CO-lazere* (Optogalvanic effect in a CO laser) (Moscow: Preprint FIAN No. 30, 1987) (in Russian).

2.44. F. O. Shimizu, K. Sasaki, K. Ueda, *Jap. J. Appl. Phys.* **22**: 1144–1151 (1983).

2.45. C. J. Walsh, *J. Phys. D.* **18**: 789–794 (1985).

2.46. E. Arimondo, M. G. Di Vito, K. Ernst, M. Inguscio, *J. de Phys.* **44**: 267–270 (1983).

2.47. E. Arimondo, *Proc. of 6 General Conf. Europ. Phys. Soc.* **1**: 354–358 (1985).

Chapter 3

Optogalvanic Spectroscopy Based on High-Frequency Electrodeless Gas Discharge

3.1 INTRODUCTORY REMARKS

The majority of the experimental and theoretical works performed to date on the optogalvanic effect have been concerned with the positive column of gas discharges, flames, and hollow-cathode discharge plasmas. However, research into the impedance response of high-frequency, usually electrodeless, gas discharges to the optical excitation of selective transitions also proves very interesting and promising for a number of reasons.

Let us consider the merits of precisely this type of discharge.

Compared to a dc-excited discharge, a stable high-frequency discharge plasma can fairly easily be sustained at substantially lower gas pressures and higher electron temperatures. This makes it easier to use the optogalvanic detection technique in conjunction with various nonlinear laser spectroscopy and degenerate-state interference methods [3.1, 3.2], to take precision linewidth and shift measurements for spectral lines starting at high Rydberg levels 3.3], and so on.

When there are no electrodes inside the discharge tube, the range of gases and gas mixtures open to investigation is extended materially, for the problems associated with the atomization and corrosion of the electrode material are in that case eliminated. It is known that high-frequency induction discharges have formed the basis of various types of electrodeless plasmatrons possessing important advantages over their arc-discharge counterparts. Specifically, they have substantially extended the possibilities for the plasma-chemical production of ultrahigh-purity compounds so important to microelectronics, sterile granules of refractory materials for fine metallurgy, etc. [3.4]. Also, the service life of electrodeless plasmatrons is much longer. Hence, it is easy to understand the importance and urgency of adopting optogalvanic spectroscopy techniques with a view to changing purposefully the characteristics of such plasmatrons.

In recent years, the electrodeless high-frequency plasmatron (the so-called inductively coupled plasma plasmatron) has actually revolutionized emission spectral analysis [3.5]. It has sharply improved the precision, reproducibility, and sensitivity of determination, extended the concentration range covered by the method, and opened up extensive possibilities of automating analytical operations.

One can also note that high-frequency discharges are helping to extend the optogalvanic method to the study of complex molecules and low-stability radicals [3.6–3.9]. Finally, emphasis should be placed on the practically important possibility of cutting down materially the contribution to the recorded optogalvanic signal from the noise component by way of frequency modulation of the discharge [3.6, 3.10].

Considering the above merits of electrodeless high-frequency discharges, certain steps have been taken along the following two lines: first, the refining and improvement of the discharge model itself, generally speaking, irrespective of the optogalvanic spectroscopy experiments [3.11, 3.12] and secondly, the development of the theory of the optogalvanic effect necessary for the electrodeless high-frequency discharge, as well as for some other types of discharge wherein purely electrodynamic specifics are not too strongly manifest [3.13–3.15]. The main results of the above works will be considered below after a brief review of the known literature data on the physics of high-frequency induction discharges.

In this chapter (Sects. 3.2 through 3.4), emphasis is placed on the model high-frequency discharge characteristics not always directly related to the optogalvanic experiments described in the literature. Nevertheless, familiarization with the existing approaches to the problem, starting with the elementary "metal cylinder" model, is necessary, for it is on this basis that the current status of the works on optogalvanic spectroscopy using electrodeless high-frequency plasmas can be evaluated.

3.2 **METAL CYLINDER MODEL**

Consider an electrodeless induction discharge (in other words, an H-type discharge, Fig. 3.1.). A high-frequency current is passed through a coil – solenoid (inductor); the alternating magnetic field of this current only has a z-component, H_z, i. e., it is directed along the axis of the coil. This field induces a vortex electric field, E_φ, which initiates and sustains a discharge in a tube made of a dielectric material. The tube is placed inside the solenoid and is filled with the gas or gas mixture of interest. If necessary, unsteady glow conditions can easily be produced by passing a strong current pulse through the inductor. In practice, the breakdown is often effected independently, for example, by means of an auxiliary electrode heated independently.

If R is the radius of the discharge tube, the plasma radius r_0 may, in

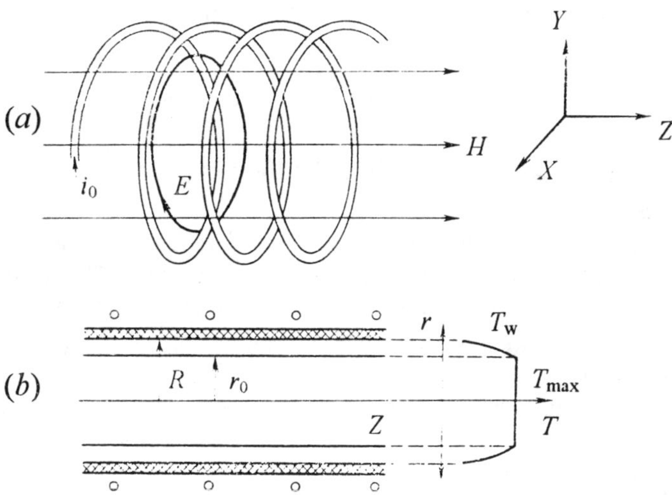

Figure 3.1. (a) Illustrating the initiation of an electrodeless high-frequency discharge; (b) cross-sectional view of the discharge tube and the general plasma temperature distribution.

principle, be smaller than R, too. In simple high-frequency discharge models, it is usually presumed that the plasma temperature $T = T_{max}$ within the limits of the radius r_0 and then drops down to the wall temperature T_w steeply enough (see Fig. 3.1b).

Disregarding the bias current and putting E, $H \sim \exp(-i\omega t)$, the Maxwell equations for the fields should be written in the form

$$\frac{1}{r}\frac{d}{dr}\left(rE_\varphi\right) = \frac{i\omega H_z}{c}, \quad -\frac{dH_z}{dr} = \frac{4\pi}{c}\sigma E_\varphi. \tag{3.1}$$

On the other hand, for the energy balance equation (the Elenbaas-Heller equation in other words), we have

$$\frac{1}{r}\frac{d}{dr}\left(k_T r\frac{dT}{dr}\right) + \sigma\left\langle E_\varphi^2\right\rangle = 0, \tag{3.2}$$

where k_T is the thermal conductivity coefficient. It is quite natural to write the boundary conditions for Eqs. (3.1) and (3.2) in the form

$$E_\varphi\left(r=0\right) = 0, \quad k_T\frac{dT}{dr}\bigg|_{r=0} = 0,$$

$$T\left(r=R\right) - T_w \cong 0, \quad H_z\left(r=R\right) \equiv H_0 = 4\pi i_0 n / c, \tag{3.3}$$

where i_0 is the strength of the current flowing through the inductor coil and n is the number of turns per unit length of the coil. The complex amplitudes of i_0 and H_0 are usually taken to be real, and the phase shifts of the fields H and E_φ are usually reckoned from the phase of the field H_0 near the inductor.

Some approximate estimates of the high-frequency induction discharge plasma characteristics can be obtained within the framework of the simple and frequently used metal cylinder model. Its essence is that the plasma column of radius r_0 is treated as a long metal rod, its conductivity σ_m being taken to be a constant quantity found independently.

A specific feature of the metal cylinder model is the possibility of splitting the electrodynamic and thermal problems in a natural fashion. Assuming that the quantities σ_m and r_0 are specified parameters, the general solution of the set of Eqs. (3.1) for the case of cylindrical conductor can be derived but it is fairly cumbersome and difficult to observe, so it is simpler to restrict oneself to the case of strong skin effect,

$$\delta \ll r_0, \quad \delta = c(2\pi\sigma\omega)^{-1/2}, \tag{3.4}$$

where the field penetrates the plasma but to a shallow depth, and the geometry of the problem can to a good approximation be considered to be plane. In that case, if one directs the x-axis into the plasma layer in a direction opposite to the radius r and the y-axis, along a tangent to the plasma surface, the origin of coordinates being selected to lie on the surface, one should write the following expressions instead of Eqs. (3.1) and (3.3):

$$\frac{dH_z}{dx} = \frac{4\pi}{c}\sigma E_y, \quad \frac{dE_y}{dx} = -\frac{i\omega}{c}H_z,$$
$$H(x=0) = H_0, \quad E_y(x\to\infty) = H_z(x\to\infty) = 0. \tag{3.5}$$

It is not very difficult to see that the above expressions are equations describing a plane electromagnetic wave in a medium with a purely imaginary dielectric constant. The solutions of Eqs. (3.5) in complex form are as follows:

$$E_y = H_0(\omega/4\pi\sigma)^{1/2}\exp(-x/\delta)\exp\left[-i\left(\omega t - x/\delta + \pi/4\right)\right],$$
$$H_z = H_0\exp(-x/\delta)\exp\left[-i\left(\omega t - x/\delta\right)\right]. \tag{3.6}$$

We have thus obtained the results that are well known in classical electrodynamics: the amplitudes of the fields E_{y0} and H_{z0} decrease exponentially as one goes deep into the plasma; the ratio $E_{y0}/H_{z0} = (\omega/4\pi\sigma)^{1/2} \ll 1$ (for the "commercial" frequency $f = 13.5$ Mhz, this ratio amounts to some 5×10^{-4}); the electric and magnetic fields are $(\pi/4)$-deg. out of phase. The electromagnetic energy flux directed into the plasma column at right angles to its surface is obviously given by

$$\langle S\rangle = S_0 \exp(-2x/\delta),$$
$$S_0 = (c/16\pi)(\omega/2\pi\sigma)^{1/2}H_0^2, \tag{3.7}$$

the Joule heat sources being concentrated in the surface layer with an effective depth of $\delta/2$.

The metal cylinder model also enables one to obtain a number of important relationships concerning the temperature characteristics of high-frequency plasmas. To illustrate, it has proved possible [3.19] to indicate, in addition to the well-known integral of the fluxes [3.18]

$$r^{m}(\mathbf{J} + \langle \mathbf{S} \rangle) = \text{const}, \tag{3.8}$$

where $\mathbf{J} = -k_T \nabla$, $\mathbf{S} = [c/(4\pi)][\mathbf{E} \times \mathbf{H}]$, $m = 0$ for a plane geometry, $m = 1$ for a cylinder, and $m = 2$ for a sphere, which is widely used in gas discharge physics, also a second (exact) integral for the set of plasma energy equations and Maxwell equations. It has the form

$$\int_{T_w}^{T_{max}} \sigma(T) k_T(T) dT = c^2 H_0^2 / 64\pi^2 = \left(i_0 n\right)^2 \tag{3.9}$$

and formally defines the temperature of the high-frequency plasma as a function of the inductor current.

The following relations are being frequently used for estimation purposes:

$$\sigma(T) = C \exp\left(-I/kT\right), \quad C(T) \cong \text{const}, \tag{3.10}$$

where I is the ionization potential of the gas. Since the relation $I \gg T_{max}$ usually holds true, one can obtain from Eq. (3.9) the following expression:

$$T_{max} \cong I/\{2[\ln\left(4k_{T\,max}C/I\right) - \ln\left(i_0 n\right)]\,k\}. \tag{3.11}$$

One can see from the above expression that to attain high T_{max} values is not easy as it requires high ampere-turn $(i_0 n)$ values and proportionately high values of the power dissipated per unit length of the cylindrical conductor: $F = 2\pi r_0 S_0$. Indeed,

$$F \cong 2\pi R S_0 \approx H_0^2 \sigma_m^{-1/2} \approx i_0 n \approx \sigma_m^{1/2}. \tag{3.12}$$

If the skin layer is thin, the temperature T_{max} is independent of the field frequency.

Summarizing, despite its interesting corollaries and applications, the metal cylinder model fails to provide any self-consistent solution of the problem for the plasma + field system; it is only partially helpful in clearing up the question of the temperature characteristics of the discharge, and is absolutely ineffective as far as the calculation of the impedance response of the discharge to a selective laser action is concerned. For these reasons, the model should be radically improved.

3.3 SELF-CONSISTENT AMBIPOLAR DIFFUSION MODEL. INITIAL EQUATIONS AND SIMPLE ESTIMATES

The well-known Schottky ambipolar diffusion model is often used in glow discharge plasma parameter calculations [3.20] and can also be used for numerical calculations of the high-frequency induction discharge plasma parameters. In this connection, reference should be made to the earlier works [3.21, 3.22] whose authors were primarily interested in the field characteristics of discharges. No self-consistent solution of the problem was, however, obtained in these works. Mention should also be made of the vast series of calculations performed in [3.23] for elecrodeless high-frequency tubes (see also the references cited therein). These calculations were also based on the ambipolar diffusion model, and their objective was to determine the temperature and level populations of the components of plasmas produced in mixtures of inert gases and metal vapors, the authors' focus being mainly low-power discharges with a poorly manifest skin layer. No electromagnetic field components were calculated in [3.23], and the energy balance was taken into account by a simplified scheme; what is more, as noted by the authors themselves, the numerical method used was not very reliable.

We have already mentioned that an attempt was made in [3.11, 3.12] at constructing such a self-consistent solution of the problem as would determine both the high-frequency discharge plasma parameters and the electromagnetic field distribution. The Schottky ambipolar diffusion theory was only used in these works as one of the possible model approaches, and the numerical techniques made it possible to formulate and solve much more general problems. Specifically, the qualitative estimates presented in the works cited above allowed the authors to establish the relationships between the energy deposited in the discharge and the mean electron density and temperature values on the one hand and the external field strength, the gas species in hand, and the geometrical factors of the problem under consideration on the other hand. The quantitative estimates were mainly made for argon discharges.

As before, we assume that the plasma is an axially homogeneous cylinder of radius r_0 placed in an inductor, and it is only the dependence of the problem parameters on the radial variable r that is retained. In the subsequent discussion, the electron velocity distribution function will be taken to be of equilibrium type:

$$f_0 = n_{\mathrm{e}}(m_{\mathrm{e}}/2\pi k T_{\mathrm{e}})^{3/2} \exp\left(-m_{\mathrm{e}}v_{\mathrm{e}}^2/2kT_{\mathrm{e}}\right). \tag{3.13}$$

The electron temperature T_{e} is constant all over the cross section of the plasma column, and the electron density n_{e} is certainly not expected to remain constant either in time or over the cross section of the column. The assumption that $T_{\mathrm{e}} = \mathrm{const}$ in Eq. (3.13) is justifiable because of the high electronic thermal conductivity at low gas densities ($p \leq 1$–10 Torr) [3.20] and is proven experimentally even within a substantially wider range of gas pressures. Note also that the electron-ion recombination practically takes place on the discharge tube walls only: at $n_{\mathrm{e}} \leq 10^{13}$ cm^{-3} bulk recombination can be disregarded (see [3.20], p. 240).

In the ambipolar diffusion model possessing the property of self-consistency in conditions of explicitly manifest skin layer, the initial system of equations describing the high-frequency discharge assumes the following form:

$$\frac{\partial n_{\mathrm{e}}}{\partial t} = D^+ \nabla^2 n_{\mathrm{e}} + S_{\mathrm{e}} n_{\mathrm{e}},$$

$$\frac{3}{2}\frac{\partial}{\partial t}\left(n_{\mathrm{e}}kT_{\mathrm{e}}\right) = \mathrm{div}\,\mathbf{q}_{\mathrm{T}} + W_+ - W_- + Q_+ - Q_-,$$

$$\frac{\partial n_m}{\partial t} = -A'_m n_m - n_m \sum_{k \neq m} S_{mk} + \sum S_{km} n_k - S'_m n_m, \tag{3.14}$$

$$\mathrm{rot}\,\mathbf{H} = \frac{4\pi\sigma}{c}\,\mathbf{E}, \quad \mathrm{div}\,\mathbf{H} = 0,$$

$$\mathrm{rot}\,\mathbf{E} = -\frac{i\omega}{c}\,\mathbf{H}, \quad \mathrm{div}\,\mathbf{E} = 0.$$

Let us clarify the notation used in the above system of equations and provide the necessary explanations.

The first equation defines the balance of the electron density $n_{\mathrm{e}}(r, t)$ in the course of ambipolar diffusion characterized by the coefficient D^+, $S_{\mathrm{e}}(r, t)$ being the rate coefficient defining the production in the plasma of fresh electrons as a result of the electron-impact ionization of atoms from the population n_k:

$$S_\text{e} = \sum n_k \left\langle \sigma_k v_\text{e} \right\rangle \exp\left(-I_k / kT_\text{e}\right). \qquad (3.15)$$

Here σ_k is the nonexponential part of the ionization cross section for the kth level, I_k is the corresponding ionization potential, v_e is the electron velocity, and the symbol $\langle ... \rangle$ denoted the operation of averaging over Maxwell distribution (3.13). Regarding the first equation in system (3.14), one should also consider the pertinent boundary conditions

$$n_\text{e}\left(R,t\right) = 0, \quad \left.\frac{\partial n_\text{e}}{\partial r}\right|_{r=0} = 0 \qquad (3.16)$$

associated with the electron recombination on the discharge tube walls.

The second equation of the system expresses the electron energy balance and needs to be explained in more detail. As distinct from equation (3.13), the electron temperature T_e in the given case should no longer be generally treated as a constant. Here the quantity

$$\mathbf{q}_\text{T} = -k_\text{T}{}^\text{e}\nabla T_\text{e} + \chi_\text{e}\varphi(T_\text{e}), \qquad (3.17)$$

whose form follows directly from the Boltzmann kinetic equation [3.24], defines the overall (conductive + convective) electronic heat-transfer flux, $k_\text{T}{}^\text{e}$ is the electronic thermal conductivity coefficient, and χ_e is the directed electron flux density; the function $\varphi(T_\text{e})$ depends on the form of the electron-by-atom elastic scattering cross section [3.24], this factor being of the same order of magnitude as T_e.

The terms W_+ and W_- describe the sources and sinks in the heating and cooling of the electronic gas as a result of elastic and inelastic electron-atom collisions, Q_+ stands for the energy deposition from the external field, and Q_- is the work done to oppose the forces of the field resulting from the ambipolar diffusion process. These four terms may be written as follows:

$$W_+ = \sum_{k=1}^{r} \sum_{m>k} E_{km} n_m n_\text{e} \left\langle \sigma_{mk} v_\text{e} \right\rangle \exp\left(-E_{km} / kT_\text{e}\right),$$

$$W_- = \sum_{k=2}^{r} \sum_{m<k} E_{km} n_m n_\text{e} \left\langle \sigma_{mk} v_\text{e} \right\rangle \exp\left(-E_{km} / kT_\text{e}\right) +$$

$$+ \sum_{k=1}^{r} I_k n_k n_e \left\langle \sigma_{mk} v_e \right\rangle \exp\left(-I_k / kT_e\right),$$

$$Q_+ = \left\langle \mathbf{jE} \right\rangle, \quad Q_- = \left\langle \mathbf{j}_a \mathbf{E}_a \right\rangle. \tag{3.18}$$

Here E_{km} is the energy difference between the levels k and m, σ_{mk} is the nonexponential part of the cross section for the collisional transition $m \to k$, $\mathbf{J}$ is the density of the current produced by the inductor, $\mathbf{j}_a$ is the density of the current due to ambipolar diffusion, and $\mathbf{E}$ and $\mathbf{E}_a$ are the respective electric field strengths corresponding to these current densities (and collinear with them).

Estimates show that at gas pressures of the order of 1 Torr in the tube the loss of energy through elastic collisions can be disregarded; this inference coincides with the data presented in [3.23].

The second equation of the system should be supplemented with the boundary condition for the temperature at the discharge tube wall. When electrons undergo wall recombination, the energy flux becomes exactly equal to the second term on the right-hand side of Eq. (3.17); in other words, the following boundary condition must hold true:

$$\left. \frac{dT_e}{dr} \right|_{r=R} = 0. \tag{3.19}$$

A series of control model calculations performed with due regard for condition (3.19) has shown that the high thermal conductivity of the electronic gas makes the electron temperature T_e practically constant all over the cross section of the discharge tube, provided that $r_0 \cong R$. In that case, the electron energy balance equation can be integrated over the volume occupied by the plasma:

$$\frac{3}{2} kT_e \frac{d}{dt} \int_V n_e dV = \int_V \left(W_+ - W_-\right) dV + \int_V \left(Q_+ - Q_-\right) dV - J_\varphi(T_e), \tag{3.20}$$

where the last term on the right-hand side is the overall electron flux onto the discharge tube wall. It is small in the case of ambipolar diffusion, and so it can, as a rule, be neglected in calculations.

Next we consider the subsystem of equations written as the third equation in system (3.14) and characterizing the balance of the atomic population n_m. In the discussion to follow, it will be quite natural to introduce an additional term defining the laser pumping of the level m

exactly into this subsystem of equations. In the present discussion, such term is absent, and so, incidentally, are a great many other terms that might prove important when the conditions of the problem are altered. But it seems reasonable to restrict ourselves to the minimum necessary set of the most important elementary processes. We will revert to the discussion of the subsytem of equations for level populations in Sect. 3.5.

The system of equations (3.14) terminates in the Maxwell equations which are simplified in an obvious manner to fit the description of the high-frequency field in conditions of explicitly manifest skin layer. In these equations, σ stands for the complex conductivity of the discharge plasma. According to [3.22], the concrete form of the conductivity is as follows:

$$\sigma = \sigma_R + i\sigma_I = \frac{n_e e^2 \left(v_t - i\omega\right)}{m_e\left(v_t + \omega^2\right)}. \tag{3.21}$$

We supplement the Maxwell equations with the boundary conditions

$$H(r = R) = H_0, \quad \left.\frac{\partial H}{\partial r}\right|_{r=0} = 0, \quad E(r = 0) = 0 \tag{3.22}$$

which have already been partially introduced when analyzing the metal cylinder model. As before, the field components H_z and E_φ are other than zero and the ratio between the electric and magnetic field amplitudes is defined by the quantity $(\omega/4\pi\sigma)^{1/2} \ll 1$.

Though the main specifics of the model described can only be revealed through numerical calculations based on the entire system of equations (3.14), one can make a number of useful general estimates, as demonstrated in [3.14].

For example, the specific high-frequency field energy deposition in the plasma, simply calculated subject to the condition $\delta \ll r_0$, is

$$Q_+^\Sigma = V^{-1} \int_V Q_+ dV \sim H_0^2 \, \bar{n}_e^{-1/2}, \tag{3.23}$$

where $\bar{n}_e$ is the electron density averaged over the cross section of the plasma column. One can also demonstrate that the overall loss of energy is linear in $\bar{n}_e$, i. e.,

$$P_-^{\Sigma} = V^{-1} \int_V \left(W_- + Q_-\right) dV \sim \overline{n}_{\mathrm{e}}, \qquad (3.24)$$

whence we get by equating Q_+^{Σ} to P_-^{Σ}

$$\overline{n}_{\mathrm{e}} \sim H_0^{4/3}, \quad P_-^{\Sigma} \sim H_0^{4/3}. \qquad (3.25)$$

From the above relations follows an important practical conclusion as to the stability of the discharge. Indeed, when the electron density $\overline{n}_{\mathrm{e}}$ fluctuates to grow higher, the depth δ of the skin layer is reduced (see (3.4) and (3.21)), the specific field energy deposition decreases in accordance with Eq. (3.23), which in turn causes $\overline{n}_{\mathrm{e}}$ to lower by virtue of Eq. (3.24) as a result of the equalization of Q_+^{Σ} and P_-^{Σ}. Thus, in accordance with the Le Chatelier-Braun principle, the plasma + field system will possess some stability margin.

It is also useful to analyze the properties of the model equations in the case of small deviations of the quantities n'_{e} and T'_{e} from the averaged equilibrium values $\overline{n}_{\mathrm{e}}$ and T_{e}. If we introduce in accordance with [3.24] the characteristic time τ_{a} for the ambipolar diffusion process, we can easily get the following linearized equations:

$$\frac{dn'_{\mathrm{e}}}{dt} = -\frac{n'_{\mathrm{e}}}{\tau_{\mathrm{a}}} + AT'_{\mathrm{e}},$$

$$\frac{dT'_{\mathrm{e}}}{dt} = -Bn'_{\mathrm{e}}, \qquad (3.26)$$

where

$$A = \frac{\partial S'_{\mathrm{e}}}{\partial T_{\mathrm{e}}} \overline{n}_{\mathrm{e}} \quad \text{and} \quad B = \left|\frac{\partial Q_+^{\Sigma}}{\partial n_{\mathrm{e}}}\right|.$$

We can see that subject to the condition

$$1/2\tau_{\mathrm{a}} - AB \geq 0 \qquad (3.27)$$

the system of equations (3.26) has a purely decaying solution; in the case of inverse inequality, oscillations occur with the frequency

$$\omega_1 = [AB - (1/2\tau_a)^2]^{1/2}. \tag{3.28}$$

The physical nature of these oscillations is associated with the Le Chatelier-Braun principle mentioned above, and the decay is due to the dissipative processes taking place in the system.

3.4 SELF-CONSISTENT AMBIPOLAR DIFFUSION MODEL. NUMERICAL CALCULATIONS

One of the approaches to the solution of the problem is based on the use of properties of either the classical Schottky model or some modification of it [3.25, 3.26]; another approach is the direct numerical solution of the system of equations (3.14).

If there are grounds to believe (in the case of not very high H_0 and n_e values) that the ionization of atoms takes place mainly from their ground state, the solution for $n_e(r)$ can at once be written in the form

$$n_e(r) = n_e(0)J_0(\xi r), \quad \xi = (S_1/D^+)^{1/2}, \tag{3.29}$$

where J_0 is a zero-order Bessel function. In that case, the boundary condition $n_e(R) = 0$ enables one to find the electron temperature T_e from the obvious relation

$$\xi R = 2.405. \tag{3.30}$$

Note that Granovsky [3.20] formulated a more exact boundary condition presupposing a nonzero (though very low) concentration of charge carriers near the discharge tube walls. It has the form

$$n_e(R)/n_e(0) \cong (\lambda_i/R)(T_e/T_i), \tag{3.31}$$

where λ_i is the ion free path length ($\lambda_i \ll R$) and T_i is the ion temperature. In the given case, one can find the ratio T_e/T_i from the equation

$$\xi R = 2.405[1 + 0.82(\lambda_i/R)(T_e/T_i)]^{-1} \tag{3.32}$$

if one uses some or other estimate of the ion free path length [3.17].

To find $n_e(0)$, use can be made of the stationary limit of integral balance equation (3.20).

Let us make several general remarks concerning the numerical technique used to solve the system of equations (3.14). Of principal physical interest is the steady-state condition of the plasma, i. e., the situation where $\partial/\partial t = 0$. However, as far as computation is concerned, it is more convenient to solve nonstationary equations subject to some initial conditions or, to put it differently, to employ the establishment procedure [3.27]. Stationary solutions are in that case obtained in the limit of high temporal variable values.

To illustrate, in discretizing the first equation of system (3.14), one can specify a uniform grid with a mesh width of h along the discharge tube radius and also take a uniform time step of τ. Putting $n_e(ih, j\tau) = n_i^j$, we approximate the ambipolar diffusion equation by the following finite-difference analog:

$$\frac{n_i^{j+1} - n_i^j}{\tau} = D_i^{+j} \frac{n_{i+1}^{j+1} - 2n_i^{j+1} + n_{i-1}^{j+1}}{h^2} + \frac{1}{R_i} \frac{n_{i+1}^{j+1} - n_{i-1}^{j+1}}{2h} + S_{ei}^j n_i^{j+1},$$
$$i = 0, 1, \ldots, N - 1, \quad j = 0, 1, 2, \ldots, \tag{3.33}$$

where D_i^{+j} and S_{ei}^j are coefficients calculated with the plasma parameters taken from the jth time layer and N is the number of grid nodes on the radius. Equation (3.33) is further solved by the fitting technique [3.28] with due regard for boundary conditions (3.16) or (3.15).

The computation procedure is as follows. Based on the given initial distribution $n_e(r)$, one determines the fields $\mathbf{E}$ and $\mathbf{H}$, the populations n_m, and the electron temperature T_e. The data obtained are then used to find D_i^{+j} and S_{ei}^j and construct the distribution $n_e(r)$ for the next instant of time by means of Eq. (3.33). Computations are repeated until the norm of the difference between the preceding and subsequent solutions becomes smaller than the preset value, thus indicating that the solution has reached its steady-state limit.

To calculate the radial distributions of H_z and E_φ, it is convenient to form from the Maxwell equations entering into system (3.14) the second-order differential equation

$$\nabla^2 H_z - i\delta^{-2} H_z = 0, \tag{3.34}$$

where δ is the depth of the skin layer, and approximate it, by analogy with Eq. (3.33), by a finite-difference analog on the same spatial grid:

$$\frac{H_{k+1} - 2H_k + H_{k-1}}{h^2} + \frac{1}{R_k}\frac{H_{k+1} - H_{k-1}}{2h} - i\delta_k^{-2}H_k = 0, \qquad (3.35)$$

$$k = 0, 1, ..., N-1, \quad H_N = H_0.$$

Once the system of equations (3.35) has been solved by the matrix sweep technique (one should bear in mind that H_z is a complex quantity), one can easily find E_φ from the Maxwell equations. One of the calculated results for the case of weakly manifest skin layer is presented in Fig. 3.2.

When calculating the populations of the excited atomic states, n_m, one solves a stationary version for the group of equations entering into system (3.14) in the form of the third equation. Of principal importance in doing so is the choice of the minimum necessary number r of levels to be taken into account (see later in the text). And finally, the last to be solved (and

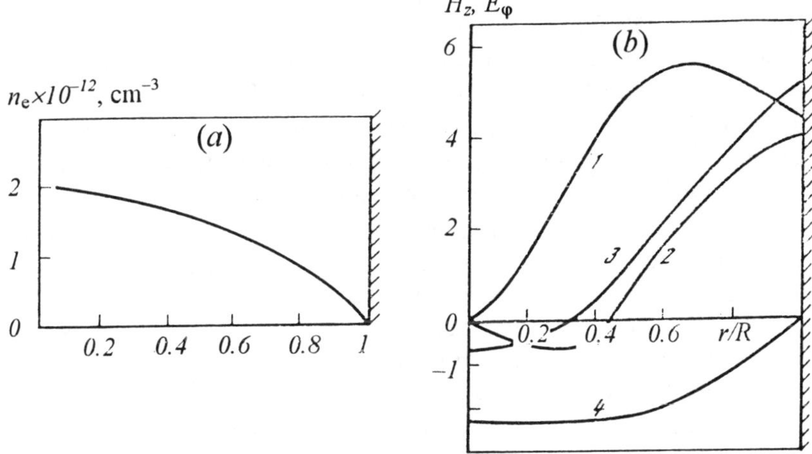

Figure 3.2. (*a*) Radial electron density profile of a high-frequency plasma; (*b*) real and imaginary components of the fields E_φ and H_z. *1* – Re(E_φ); *2* – Im(E_φ); *3* – Re(H_z); *4* – Im(H_z). Ar atom concentration $n_a = 10^{15}$ cm^{-3}, $H_0 = 5$ Oe, $R = 1.5$ cm, $T_e = 2.7$ eV (calculated), H is measured in oersteds and E_φ, in rot H.

again in the steady-state limit) is Eq. (3.20) for the unknown electron temperature. The integrals entering into Eq. (3.20) are approximated by the Simpson formula, and the solution itself is found by the Newton method.

As previously noted, concrete calculations were most frequently performed for high-frequency discharges in pure argon where closely spaced energy levels were united into blocks in accordance with the recommendations and criteria presented in [3.29]. Use was made of the oscillator force data borrowed from [3.30] and, when calculating the cross sections of inelastic processes, the empirical Dravin formulas also presented in [3.29].

A fairly large series of numerical calculations were performed with a view to finding the characteristic parameters of electrodeless high-frequency discharges and verifying some general corollaries of the suggested self-consistent model illustrated in [3.11, 3.12]. The authors aimed primarily to solve the question as to the methods of modeling the atomic ionization mechanism. According to the Schottky model, ionization predominantly proceeds from the ground state. In the modifications of this model described in [3.25, 3.26, 3.31], ionization from the excited levels is also taken into consideration, but it is assumed that the rate coefficient $S_e =$ const, i. e., it is taken to be independent of the radial variable. In that case, expressions (3.29) and (3.30) hold true as with the original Schottky model. It should be emphasized that the possibility of using these expressions substantially eases the problem, for the first equation of system (3.14) no longer needs to be solved, and the temperature T_e can at once be determined by means of the first boundary condition (3.16) while retaining the steady-state limit of Eq. (3.20) to find $n_e(0)$.

When finding the self-consistent solution of the problem, both in its complete formulation and with the use of a modified Schottky model, the choice of the number of atomic states to be taken into account proves very important. A general inference that can be drawn in this connection is that the highest level to be considered must lie higher on the energy scale than the so-called bottleneck [3.29, 3.32] which can roughly be estimated by the relation $E_m \cong 3kT_e/2$. To verify this conclusion, the number of atomic levels taken into consideration in the numerical experiments was usually varied.

Figure 3.3 presents the results of calculations of the relationships between the maximum electron concentration $n_e(0)$ and electron temperature T_e on the one hand and the atomic gas density n_a and the strength amplitude of the external magnetic field H_0 on the other hand. As

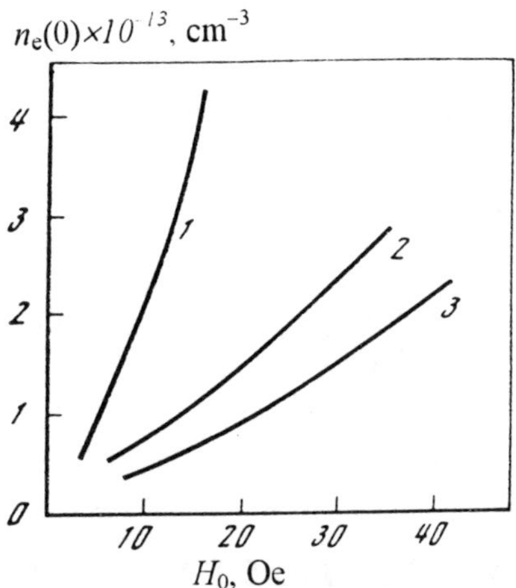

Figure 3.3. Electron concentration at the discharge tube axis as a function of the strength amplitude of the field H_0. $1 - n_a = 5 \times 10^{15}$ cm^{-3}, $T_e = 1.7$ eV; $2 - n_a = 10^{15}$ cm^{-3}, $T_e = 2.7$ eV; $3 - n_a = 7 \times 10^{14}$ cm^{-3}, $T_e = 3$ eV.

follows from the properties of the Schottky model, be it simple or modified, the temperature T_e is independent of H_0 and decreases with rising gas density. The electron concentration at the discharge tube axis in its turn grows higher with the increasing strength of the field H_0 (at a fixed n_a value) and rising n_a (at a fixed H_0 value). The numerical calculation results closely agree with the first of estimates (3.25).

The authors of [3.11, 3.12] also demonstrated and discussed a series of other calculation results concerning the dependence of the electron temperature T_e on the field characteristics of the plasma, with the proportionality $P_-^{\Sigma} \sim H_0^{4/3}$ being confirmed and different numbers of atomic levels taken into account in a modified Schottky model. Note that the calculations made in accordance with this model and the comprehensive scheme usually took account of 10-15 blocks of energy levels in the inert gas atom in hand, so that the levels lying above the sixth one could be considered hydrogen-like. At the same time, it also turned out that with sufficiently strong external magnetic fields ($H_0 \geq 100$ Oe, $n_a = 10^{15}$ cm^{-3}, $R = 1.5$ cm), the calculation results for ten- and two-level models were practically the same. So, where the fields H_0 are not very strong (≤ 10 Oe), the simple Schottky model proves quite applicable, i. e., it is sufficient to restrict oneself to a single lower level, and where $H_0 \geq 100$ Oe, the so-called instantaneous ionization model [3.29] holds true. This is confirmed by the calculations of the function $T_e(H_0)$ reaching such an asymptote as corresponds to the ionization potential I_k reckoned from the first excited level of the argon atom.

Finally, Fig. 3.4 shows the comparison between the results of model calculations performed by the complete scheme and the data of the experiment [3.33] wherein subject to measurement was the integral radiative loss in an electrodeless high-frequency discharge tube filled with argon and krypton. The close agreement between the experimental and calculation results supports the applicability of the model described.

3.5 OPTOGALVANIC EFFECT IN HIGH-FREQUENCY DISCHARGES

Let us return to the system of equations (3.14) and focus on the equations for the level populations n_m that have already been discussed in brief. We now write the population balance condition in the form

$$
\frac{\partial n_m}{\partial t} = -n_m \left(\sum_{k<m} A_{mk}\gamma_{mk} + \sum_{k \neq m} S_{mk} + S_m \right)
$$

$$
+ \left(\sum_{k>m} A_{km}\gamma_{km} n_k + \sum_{k \neq m} S_{km} n_k \right) - F\sigma_\varphi \left(n_m - \frac{g_k}{g_m} n_k \right). \tag{3.36}
$$

Compared to the former expression, the above equation allows not only for spontaneous radiative deactivation of the level m, but also for the additional population of this level as a result of radiative transitions from various levels $k > m$. The role of the radiation trapping effect is taken into account in Eq. (3.36) by means of the simple procedure [3.34] described in Sect. 2.1.

The new term appearing in Eq. (3.36) describes the contribution from the optical (laser) pumping on the transition $k \rightarrow m$. The role played by this term can be easily and naturally analyzed by means of an elementary three-level atomic model including the ground level (No. 1), a single excited(resonance) level (No. 2), and the ionization level (No. 3) [3.35]. In this model, the laser illumination of the discharge always causes the

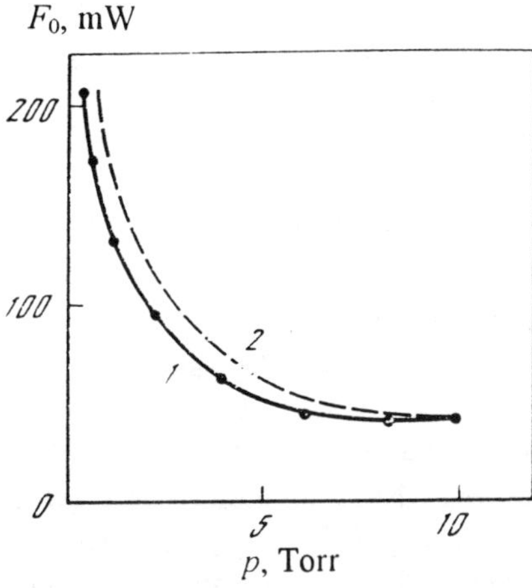

Figure 3.4. Overall radiative loss F_0 in a Bell-Blum-type high-frequency discharge tube as a function of the gas pressure (Ar, Kr) in the range 0.5–10 Torr. Discharge power $Q = 7$ W, tube radius $R = 6.5$ mm. *1* – experiment [3.33]; *2* – calculation by complete scheme (3.14).

population of the resonance level to increase, and so facilitates the growth of the electron concentration n_e and gives rise to a definite (negative) sign of the optogalvanic signal.

Based on a three-level atomic model similar to the one described in [3.35], the authors of [3.13, 3.36] numerically calculated the characteristics of the optogalvanic signal and the radial distributions $T_e(r)$, $n_e(r)$, and $H_z(r)$ for various laser excitation conditions. They constructed an implicit difference scheme of second order of accuracy in the spatial variables. As for the time variable, a scheme of first order of accuracy proved adequate, use being made of the establishment procedure [3.27]. The system of algebraic equations obtained was solved by the sweep method on a Model БЭСМ-6 computer.

No attempt was made in these calculations to reproduce the conditions of any concrete optogalvanic experiment involving a high-frequency discharge, for the energy level diagram used was obviously simplified too much. But it seemed quite appropriate to use for estimation purposes some typical initial values, for example, around 10^{-15} cm^2 for the cross section of inelastic processes, some 10 eV for the atomic ionization potential, about 3 or 5 eV for the resonance level excitation energy, 1 cm or so for the discharge tube radius, and so on. This done, it was not very difficult to follow the specific changes of both the space resolved plasma parameters and the integrated quantities governing the optogalvanic signal. An example of such a calculation is presented in Fig. 3.5 wherein the radial distributions $n_e(r)$ and $T_e(r)$ in the discharge plasma not subject to laser illumination are compared with the same distributions in the case of resonance laser irradiation under excitation saturation conditions. As one would expect, the rise of n_e corresponds to the reduction of the electron temperature T_e, the rather weakly manifest maximum of T_e at the periphery of the tube remaining in practically the same position on the axis of coordinates.

Figure 3.6. shows the relationship between $\Delta Q'/Q'$ and the pumping intensity (I^s characterizes the saturation level), where $\Delta Q'$ is the experimentally measured change of the specific energy deposition in the high-frequency discharge, which is a measure of the optogalvanic effect (see [3.6–3.9]. The initial section of this curve is scaled up in Fig. 3.6b to demonstrate its practically exact linearity with the pumping intensity; the presence of such a section was also confirmed analytically in [3.14].

The simple three-level atomic model can also be extended to a some-

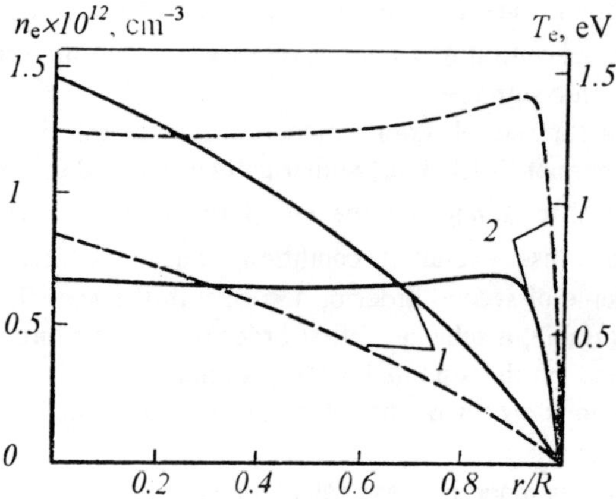

Figure 3.5. Radial distributions of (1) electron density and (2) electron temperature in the plasma of a high-frequency discharge. The solid curves correspond to the laser pumping of the plasma under excitation saturation conditions and the dashed ones are plotted in the absence of the pumping.

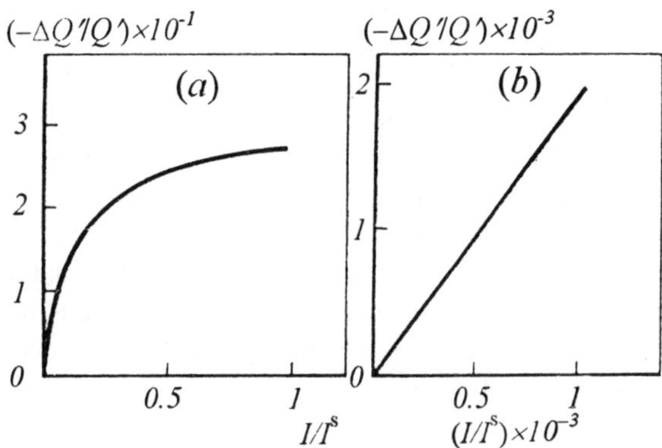

Figure 3.6. (*a*) Optogalvanic signal as a function of the pumping intensity and (*b*) the same for the initial stage of the signal rise.

what different experimental situation where the gas cell is placed in a strong magnetic field and the high-frequency energy is fed into the cell from a special generator by means of offset electrodes [3.37]. Experiments conducted with such an experimental setup hold much promise from the standpoint of studying the structure of Rydberg and autoionization states, diamagnetism of conduction electrons, and other fine effects [3.38].

Worthy of notice in this connection is also the work by Berglind and Casparsson [3.39], wherein the optogalvanic effect was observed in a flame by measuring the attenuation of a directed high-frequency field (a Gunn-effect diode, power around 10 mW, frequency around 10 Ghz) and not impedance variations as is traditionally the case. The attenuation mechanism consists in the dissipation of the deposited energy by charged particles accelerated by the field; the production of these particles is in turn initiated by laser pumping.

Let us now proceed to the results of some estimations and quantitative calculations of the optogalvanic effect characteristics in electrodeless high-frequency discharges, obtained with the aid of a multi-level atomic system model. We will first of all dwell upon the Japanese works [3.6, 3.8].

Suzuki [3.6] studied the optogalvanic effect by means of a high-frequency discharge in pure argon, the experimental conditions being such that the metastable states played no perceptible part. When performing the appropriate calculations, the authors of [3.15, 3.40] constructed a block level structure of the argon atom, with due regard for the recommendations given in [3.29], such that the local thermodynamic equilibrium conditions were presumed to be satisfied within each block. The laser pumping corresponded to the transitions among the levels of the third and sixth blocks. From 10 to 15 levels were taken into consideration, all levels starting with the sixth one being assumed to be hydrogen-like. The oscillator forces f_{km} and statistical weights g_k of the levels were borrowed from the tables presented in [3.41], and the Dravin formulas (see [3.29]) were used as a reference in calculating the cross sections of collision-induced transitions. If the pumping conditions are far from saturation, one can easily satisfy oneself that the changes in the populations of levels not belonging in the sixth block are small enough, so that the change of the high-frequency discharge power is

$$\Delta Q' \sim \lambda_{km} f_{km} g_k, \tag{3.37}$$

where k is the serial number of the level from which the pumping process proceeds. The laser pumping in [3.6] was effected at the following three wavelengths:

$$
\begin{aligned}
\lambda_1 &= 621.2 \text{ nm} & (f_1 &= 0.0033, g_1 = 5), \\
\lambda_2 &= 603.2 \text{ nm} & (f_2 &= 0.0173, g_2 = 7), \\
\lambda_3 &= 588.8 \text{ nm} & (f_3 &= 0.005, \ g_3 = 7).
\end{aligned}
$$

According to [3.41], this yields

$$\Delta Q'_1 : \Delta Q'_2 : \Delta Q'_3 = 1 : 7 : 2, \tag{3.38}$$

where the value of the quantity $\Delta Q'_1$ corresponding to the wavelength λ_1 was taken to be equal to unity. Figure 3.7 shows a simplified energy level diagram of the argon atom and the results of comparison between the experiment performed in [3.6] and estimate (3.38)[1].

There are sufficient grounds to consider these results as quite satisfactory.

As can be seen, estimation made on the basis of expression (3.37) is absolutely elementary and, in essence, not associated with the above comprehensive high-frequency discharge model providing for self-consistency. One of the results obtained on the basis of numerical

---

[1] The experimental data are presented with due regard for a somewhat different (refined) optogalvanic signal calibration that the authors of [3.40] were kindly informed of by Dr. T. Suzuki.

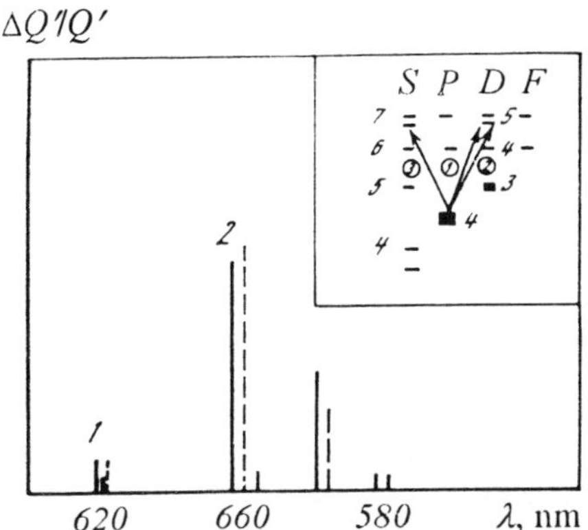

Figure 3.7. Comparison between theoretical data (dashed lines) and the results of the experiment [3.6]. $\lambda_1 = 621.2$ nm ($4p_{12} \to 5d_{23}$); $\lambda_2 = 603.2$ nm ($4p_{22} \to 5d_{33}$); $\lambda_3 = 588.8$ nm ($4p_{23} \to 7s_{12}$). The theoretical lines are shown slightly shifted to the right to make the figure more graphic.

calculations involving the entire system of equations (3.14), (3.36) is presented in Fig. 3.8. The laser pumping intensity here was taken to be higher than in [3.6]: $I = 0.25 \times 10^2$ W, n_a 5×10^{14} cm^{-3}, and the strength of the magnetic field H_0 was varied from 5 to 60 Oe. These calculations revealed that when pumping the most intense transition ($4p_{22} \to 5d_{33}$) the relative change of the electron density, $\Delta n_e/n_e^0$, could be as great as 50%. The optogalvanic signal ratios $\Delta Q'_1 : \Delta Q'_2 : \Delta Q'_3$ could vary over a fairly wide range as the field H_0 was changed. But in that case, the signal suffered no change of sign. These conclusions have been confirmed qualitatively by the results of other experiments conducted by the research group headed by Dr. T. Suzuki, but no quantitative comparison has thus far been made on the basis of the available literature data.

It is interesting to verify the feasibility of estimating the optogalvanic signal characteristics by means of a purely spectral relationship similar to (3.37), i. e., without using any model parameters specific to some or other type of discharge. Such verification was made in [3.40] regarding the very elegant experiment described in [3.42], where the discharge took place in a cesium vapor and pumping was effected in the $7S \to nP$ transitions. An IR laser was used to raise the cesium atoms to the Rydberg states nP ($n = 12$ to 18) whence they were ionized as a result of the Mölner-Hornbeck associative process possessing a great cross section, this being detected by

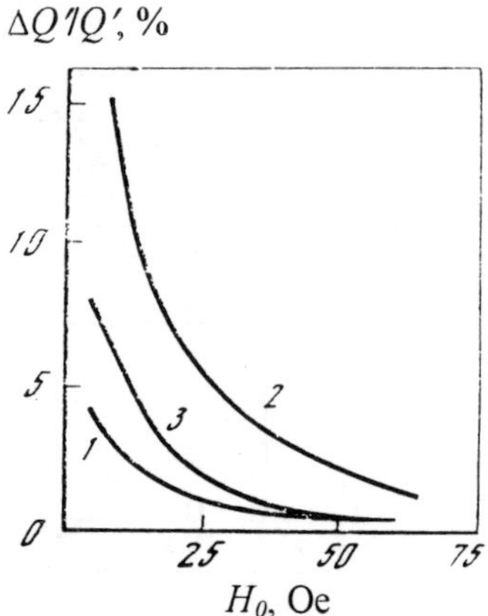

Figure 3.8. Theoretical extrapolation of the experimental data presented in [3.44] to various H_0 field strength values in oersteds. The curves 1, 2, and 3 correspond to the same spectral lines as in Fig. 3.7. $I = 25$ W, $n_a = 5 \times 10^{14}$ cm^{-3}.

means of a thermionic detector. Note that so fine an optical control of the population of the $7S$ state in cesium is of great physical interest, for it has a direct relationship to experiments aimed at revealing parity violation by means of the strongly forbidden magnetic dipole transition $6S \rightarrow 7S$ in Cs [3.43].

By analogy with Fig. 3.7, Fig. 3.9 presents the relative change of the optogalvanic signal in conditions of laser pumping in the $7S \rightarrow nP$ transitions as a function of the number n; the experimental data of [3.42] are compared with the results of calculations based again on relationship (3.37). The magnitude of the optogalvanic signal at $n = 18$ is taken to be unity. The agreement between theory and experiment can again be considered quite acceptable.

The series of experimental studies described in [3.8] were an important development of the work [3.6] considered above. In these experiments, the optogalvanic signal was detected by the double-probe technique [3.45] widely used in the practical diagnostics of inhomogeneous plasmas. This technique enabled the authors of [3.8] to improve the detection sensitivity by almost two orders of magnitude in comparison with that previously attained by them with the technique involving the measurement of the reflected high-frequency field power. Moreover, optogalvanic signals could be detected with the gas pressure varied over wide limits, for example,

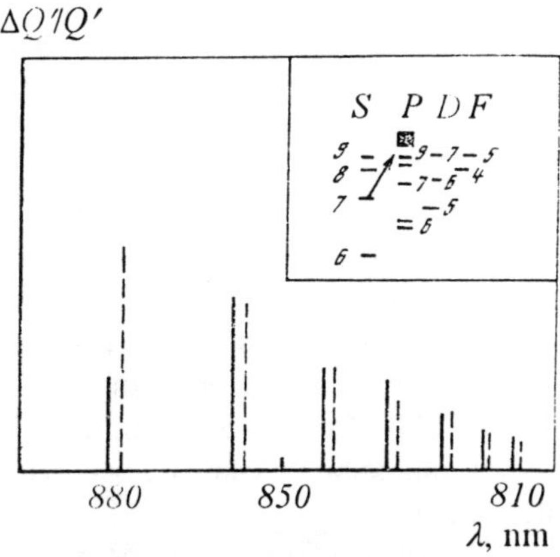

Figure 3.9. Comparison between theoretical data (dashed lines) and experimental results [3.42] (solid lines). Pumping on the transitions $7S \rightarrow nP$ (n = 12 to 18) in the cesium atom.

from 1 to 40 Torr in a helium discharge, by recording the H_2 excimer spectra with a high spatial resolution.

The well-known theoretical double-probe model [3.45] allows one to interpret the optogalvanic signal as the difference between two "floating" potential values, $\left(V_{f_1} - V_{f_2}\right)$, where

$$V_f = \frac{kT_e}{e}\left[\ln\left(\frac{n_e}{n_{e_0}}\right) - \frac{1}{2}\ln\left(\frac{\pi}{2}\frac{m_e}{M_i}\right)\right], \qquad (3.39)$$

where T_e and n_e are the local electron temperature and density, respectively, n_{e0} corresponds to the zero space potential, and m_e and M_i are the electronic and ionic masses, respectively.

It is very difficult to calculate the optogalvanic signal using the comprehensive self-consistent model based on the system of equations (3.14) because the problem loses its axial symmetry to become, at best, two-dimensional in the space variables. Nevertheless, the main corollaries concerning the laws governing the variations of n_e and T_e in high-frequency discharges, which have been discussed in Sects. 3.3 and 3.4, make it possible to estimate accurately enough the variation character of the difference ($V_{f1} - V_{f2}$) found from Eq. (3.39). To illustrate, when joining the experimentally measured and calculated values of the optogalvanic signal amplitude A_V and phase Φ_V at some point near the surface of the unearthed probe, one can monitor the behavior of these characteristics as one draws closer to the second probe (Fig. 3.10).

Note that the authors of [3.8] also described another double-probe measurement scheme wherein the probes were made in the form of two coaxial rings spaced apart along the discharge tube axis. In that case, the problem can be considered to be unidimensional to a good approximation, so that calculations by formulas (3.14) and (3.39) can be made consistently enough. Unfortunately, the worsening of the signal/noise ratio practically prevented the authors of [3.8] from experimenting further with double probes, so that it is as yet impossible to compare between experimental and theoretical data in more detail.

Let us consider the problem of using the optogalvanic spectroscopy technique with a view to improving our analytical capabilities and studying the physical properties of nonequilibrium inductively coupled plasmas that are playing, as indicated at the beginning of this section, a major part in all the works dealing with the modern emission spectral analysis. In the 80's, several works [3.46–3.49] were published on the modeling of the physical processes occurring in inductively coupled plasmas. But each of these works has strong restrictions which prevent the reconstruction of the

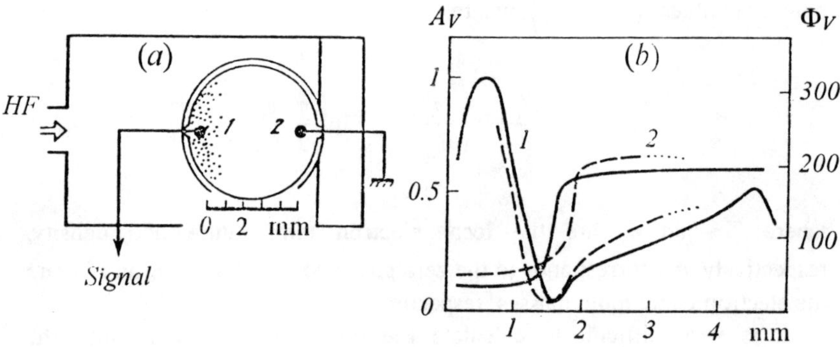

Figure 3.10. (*a*) Measuring the optogalvanic signal by the double-probe technique [3.8] and (*b*) amplitude (*1*) and phase (*2*) characteristics of the signal at various distances from the surface of the first probe. The solid curves – experimental data of [3.8] and the dashed curves – calculation results by Eq. (3.39). The amplitude A_V is measured in arbitrary units and the phase Φ_V, in degrees.

excitation, ionization, and emission mechanisms active in such plasmas; hence it is impossible to develop any acceptable theory of the optogalvanic effect therein. For example, the purely diffusive model presented in [3.46] fails to take into account a number of essential terms in the heat conduction equation, the plasma being assumed to have a double-temperature character; the possibility of any spatial inhomogeneity of the plasma flare is completely ignored in [3.47–3.49], and so on.

The experimental work by Turk and Watters [3.50] stands out among the few publications on laser photoionization of atoms in inductively coupled plasmas. This work cannot be regarded as a full-blown investigation, for, as emphasized by the authors themselves, the inductively coupled plasma parameters and laser excitation region selected, the impedance variation measurement technique used, etc. were far from being optimal. Nevertheless, even under those unfavorable conditions, the authors of [3.50] managed to obtain a number of very interesting results associated with the spectroscopy of Rydberg states, optogalvanic spectra of two-photon transitions, the role of the collisional and radiative ionization of Cu, Fe, Mn, and Na atoms, etc.

The theoretical estimates made by Bulyshev, Denisova, and Preobra-zhensky [3.15] for the characteristics of optogalvanic signals in inductively coupled plasmas as compared with the corresponding signals detected in air-acetylene and air-hydrogen flames (see [3.51, 3.52]) show the following. If atoms in a flame are excited by means of laser radiation to rise from the ground state to a level with an energy of, say, 4 eV, the rate of their ionization as a result of collisions with N_2 or O_2 molecules increases by some 7–8 orders of magnitude. Hence the extremely high sensitivity of the optogalvanic detection of impurities in flames, reaching, for example, in the case of Na, Li, and Na, around 1–3 pg/ml, i. e., approximately 10^5 atoms/cm^3 [3.51, 3.52]. If a similar photoexcitation by means of UV radiation is effected in an inductively coupled plasma, the situation will be absolutely different qualitatively. Indeed, the principal contribution to the ionization of the impurity atoms will in that case come from their collisions with the metastable argon atoms (i. e, the Penning process for energies of 11.55 and 11.72 eV [50]), as well as with excited argon atoms with an energy in the range 14–15 eV. It is evident that the relative role of the laser photoexcitation of the impurity atoms is incommensurably smaller here than in the case of flame.

However, the conclusion that the analytical optogalvanic spectroscopy based on the inductively coupled plasma is inefficient would obviously be premature and, most likely, wrong, for the experiment [3.50] itself explicitly argues against it (reliable recording of very weak lines corresponding to two-photon transitions, nonlinear character of the relationship between the laser pump intensity and the optogalvanic signal magnitude, etc.). First, it seems necessary to perform optimization works similar to [3.53, 3.54] and extend the applicability of the high-frequency plasma model described above primarily to a wider range of pressures and with due consideration for the gas dynamics of the plasma jet.

As to the optogalvanic effect in high-frequency discharges, it is interesting to discuss one possibility associated with the development of an e. m. f. in the boundary plasma layer. Branderberger [3.55] conducted his research on the experimental setup shown schematically in Fig. 3.11. The discharge tube used was a standard gas-discharge spectral light source tube filled with krypton at a pressure of around 5 Torr and equipped with internal electrodes made of tungsten. The electrodes are loaded into a differential oscillograph with a recorder 1 MΩ in input impedance. The tube envelope is embraced by two external ring electrodes connected to the output of a high-frequency oscillator. The peak output voltage of the oscillator reaches 30 V, its frequency being as high as 30 Mhz. The high-frequency discharge current reaches 1 mA and the strength of the high-

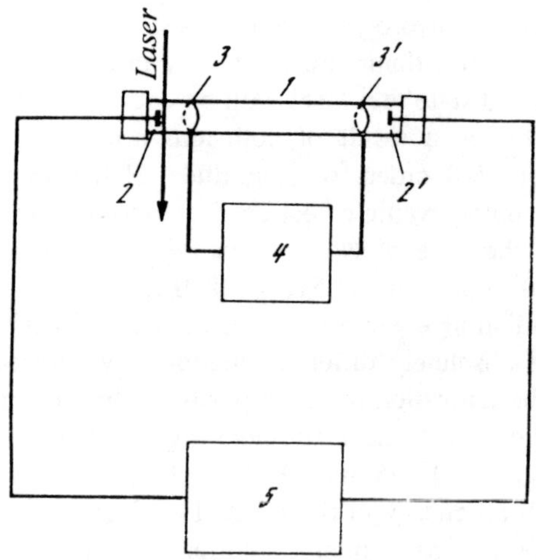

Figure 3.11. Schematic diagram of the experimental setup for e. m. f. observations [3.55]. *1* – discharge tube; *2, 2'* – internal metal electrodes; *3, 3'* – external ring electrodes; *4* – high-frequency oscillator; *5* – oscillograph with a recorder.

frequency field inside the tube, 20 V/cm. The total length of the tube is about 50 mm and the distance between the ring electrodes equals around 30 mm. The luminous zone of the high-frequency discharge has a cigar-like shape and is surrounded by a dark layer approximately 2 mm thick at the tube walls. At low values of the high-frequency discharge power, the discharge occupies approximately one half of the space between the ring electrodes, and when the discharge power values are high, it reaches as far as the metal electrodes.

It is well known from gas-discharge physics [3.16] that the inside surface of the glass discharge tube accumulates a negative charge as a result of ambipolar diffusion. This is due to the following two circumstances: (a) under steady-state conditions, the time-averaged current flowing to the dielectric wall must be equal to zero and (b) the negative surface charge repels electrons possessing a high mobility. The dark layer is occupied by positive ions and neutral atoms. As a result, in the layer near the tube walls a radial electric field exists, whose strength is comparable with that of the field in the discharge.

The situation changes near the surface of the electrically conductive internal metal electrodes which can provide for the drain of charges. But if the discharge tube is symmetrical, connecting the electrodes by an external circuit will not cause an electric current to flow in it, even if the high-frequency discharge plasma is present. However, this symmetry can be disturbed. One possible way to do so is to exert influence on the density of neutral metastable atoms which may give rise to electron emission upon their collision with the metal electrode surfaces in the course of their own deactivation (see also Sect. 2.5). A change in the density of metastable states can be caused by radiation. If, for example, the radiation reduces the density of such states in the vicinity of one of the electrodes, this will lead to a decrease in the flux of electrons injected by the electrode surface into the discharge zone, and an electric current develops as a result.

Thus, the irradiation of the electrode regions in the scheme presented in Fig. 3.11 gives rise to an e. m. f. and current without any source in the external circuit.

The work [3.55] has demonstrated the destruction of metastable levels in krypton upon its discharge being irradiated normal to its axis with a CW semiconductor laser tunable in the region of the $1s_3 \rightarrow 2p_3$ transition (785.5 nm). The radiative decay of the $2p_3$ state terminates partially (branching ratio 0.36) in the state $1s_2$ coupled by the resonance transition to the ground state of the atom. Figure 3.12 presents exemplary records of an absorption line profile whose structure is due to the presence of the ^{82}Kr, ^{83}Kr, ^{84}Kr, and ^{86}Kr isotopes and the hyperfine splitting of the odd

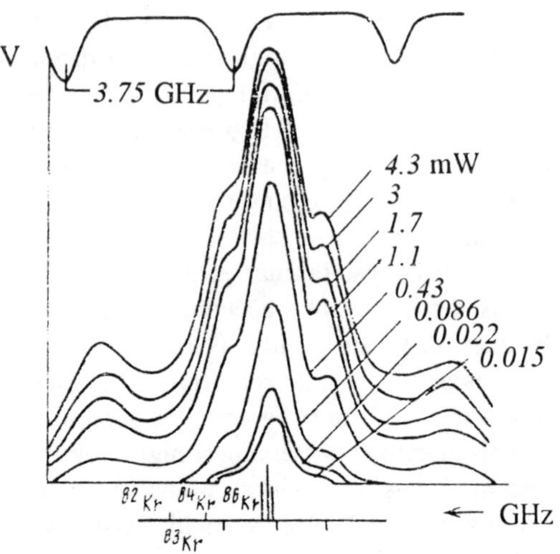

Figure 3.12. E. M. F. (in volts) as a function of the lasing frequency ν varied in the neighborhood of 785.5 nm [3.55]. The individual records correspond to different laser radiation powers. Shown at the top is the transmission record of a Fabry-Perot etalon.

isotope ^{83}Kr. The individual records correspond to different powers of the laser radiation illuminating the discharge near one of the internal metal electrodes. In this case, the resolution is limited by the Doppler broadening and the insufficient monochromaticity of the laser radiation. Specifically, the extreme left- and right-hand peaks are due to the presence of additional (side) frequencies in the lasing spectrum. The top of the figure shows the transmission spectrum of a Fabry-Perot etalon, which serves as a wavelength-scale reference. The magnitude of the voltage developing across the internal electrodes depends on the lasing power, the location of the laser beam in relation to the internal electrode, and the power of the high-frequency discharge. When the discharge is illuminated in the immediate vicinity of the electrode, the voltage typically amounts to some 100 mV at a laser radiation power of around 1 mW.

In conditions of the above experiment, the absorption of 1 μW of the laser radiation power (at a lasing power of 2 mW) causes a current of about 0.1 μA to flow in the external circuit. The density of the Kr atoms in the $1s_3$ state is estimated from the magnitude of absorption to equal around 10^{13} cm^{-3}. Based on the notions of the e. m. f. development mechanism described above and the known radiative transition characteristics, the authors of [3.55] have shown that the above parameters correspond to some 50-% probability of deactivation of Kr ($1s_3$) states accompanied by the emission of electrons.

Experiments with the discharge being illuminated at various distances from the internal electrode show that the magnitude of the e. m f. decreases as the laser beam is moved farther from the electrode and goes to zero when the beam is made to pass exactly midway between the internal electrodes.

The chief value of this method of observing the optogalvanic effect in high-frequency discharges is the fact that in the absence of irradiation of the discharge the current in the external circuit is zero, as distinct from the ordinary schemes with an external current source, wherein current variations are similar to relatively weak perturbations.

In the present chapter, we have mainly dwelt upon the optogalvanic effect in inductive high-frequency discharges, for the models described above allow for quantitative interpretation of light-induced changes in the plasma impedance. It should be noted that considerable advances have recently been made in the physics of capacitive high-frequency discharges which play an important role in plasma-chemical technologies and laser engineering. One can get an idea of the current status of this problem from the book [3.56]. The possibilities of numerically modeling such discharges as applied to active laser media were described in [3.57]. However, optogalvanic effects in such plasmas are as yet poorly known, though their study is of physical interest, particularly in connection with the specific features of their spatial structure due to the presence of the regions of a near-wall space charge devoid of electronic conductivity of its own.

REFERENCES

3.1. C. Stanciulescu, R. C. Bobulescu, A. Surmeian, *Appl. Phys. Lett.* **37**: 888–890 (1980).

3.2. P. Hannaford, D. S. Gough, G. W. Series, *J. Phys.* **44**, C7: 107–115 (1983).

3.3. E. Giacobino, F. Biraben, P. Labastie, *J. Phys.* **44**, C7: 505–511 (1983).

3.4. V. D. Rusanov, A. A. Fridman, *Fizika khimicheski aktivnoi plazmy* (Physics of chemically active plasma) (Moscow: Nauka, 1984) (in Russian).

3.5. M. Thomson, J. N. Walsh, *A Handbook of Inductively Coupled Plasma Spectroscopy* (N. Y.: Chapman & Hall, 1983).

3.6. T. Suzuki, *Opt. Commun.* **38**: 364–368 (1981).

3.7. R. Vasudev, R. N. Zare, *J. Chem. Phys.* **76**: 5267–5270 (1982).

3.8. T. Suzuki, H. Sekiguchi, T. Kasuya, *J. Phys.* **44**, C7: 419–427 (1983).

3.9. T. Suzuki, T. Kasuya, *Phys. Rev.* **36**: 2129–2133 (1987).

3.10. P. Camus, *J. Phys.* **44**, C7: 87–106 (1983).

3.11. A. E. Bulyshev, N.V. Denisova, N.G. Preobrazhensky, A.E. Suvorov, in: *Matematicheskoye modelirovaniye bezelektrodnogo vysokochastotnogo razryada* (Riga: P. Stuchka Latvian University Press, 1987): 123–132 (in Russian).

3.12. A. E. Bulyshev, N.V. Denisova, N.G. Preobrazhensky, A.E. Suvorov, *PMTF* No. 2: 3–8 (1988).

3.13. A. E. Bulyshev, N. G. Preobrazhensky, in: *Contributed Papers of XII SPIG '84. Sibenik, Yugoslavia* (1984): 481–484.

3.14. A. E. Bulyshev, N. G. Preobrazhensky, *DAN SSSR* **279**: 1357–1359 (1984).

3.15. A. E. Bulyshev, N. V. Denisova, N. G. Preobrazhensky, *Contributed Papers of XVIII ICPIG. Swansea, England* (1987), Vol. 4: 762–763.

3.16. Yu. P. Raizer, *Osnovy sovremennoi fiziki gazorazryadnykh protsessov* (Fundamentals of the modern physics of gas-discharge processes) (Moscow: Nauka, 1980).

3.17. B. E. Cherrington, *Gaseous Electronics and Gas Lasers* (Oxford, New York: Pergamon Press, 1982).

3.18. Yu. P. Raizer, *Fizika gazovogo razryada* (Gas discharge physics) (Moscow: Nauka, 1987) (in Russian).

3.19. V. A. Gruzdev, R. E. Rovinsky, A. P. Sobolev, *PMTF* No. 1: 143-150 (1967).

3.20. V. L. Granovsky, *Elektrichesky tok v gaze* (Electric current in gases) (Moscow: Nauka, 1971) (in Russian).

3.21. D. R. Keefer, *AIAA Paper* No. 703: 83–94 (1969).

3.22. B. B. Henriksen, D. R. Keefer, M. N. Clarkson, *J. Appl. Phys.* **42**: 5460–5464 (1971).

3.23. A. S. Agapov, A. A. Matveyev, V.I. Khutorshchikov, in: *Metody raschyota parametrov VCh-bezelektrodnykh lamp* (Methods for calculating the parameters of HF electrodeless tubes) (Riga: Latvian University Press, 1985): 89–98 (in Russian).

3.24. E.M. Lifshits, L.P. Pitayevsky, *Fizicheskaya kinetika* (Physical kinetics) (Moscow: Nauka, 1979) (in Russian).

3.25. H. B. Valentini, *Opt. Commun.* **53**: 313–318 (1985).

3.26. H. B. Valentini, N. G. Preobrazhensky, in: *Kineticheskaya model gazorazryadnogo optogalvanicheskogo effekta* (Kinetic model of the gas-discharge optogalvanic effect) (Novosibirsk: ITPM SO AN, 1968) (in Russian).

3.27. S. K. Godunov, *Metody matematicheskoi fiziki*) (Mathematical physics techniques) (Moscow: Nauka, 1974) (in Russian).

3.28. D. Potter, *Computational Physics* (New York: Wiley, 1973).

3.29. L. M. Biberman, V. S. Vorobiyev, I. T. Yakubov, *Kinetika neravnovesnoi nizkotemperaturnoi plazmy* (Noenequilibrium low-temperature plasma kinetics) (Moscow: Nauka, 1982) (in Russian).

3.30. W. L. Wiese, N. W. Smith, B. M. Glennon, *NBS Rept. NSRDS* **4**: 187 (1966).

3.31. N. N Zaitsev, N. Ya. Shaparev, *Optoelektricheskiye yavleniya v plazme* (Optoelectric phenomena in plasma) (Krasnoyarsk; Preprint IF SO AN SSSR No. 207, 1982) (in Russian).

3.32. S. Byron, R. C. Stabler, P. I. Bortz, *Phys. Rev. Lett.* **8**: 376–381 (1962).

3.33. S. V. Semenov, G. M. Smirnova, V. I. Khutorshchikov, *Voprosy Radioelektroniki. Ser. OVR* **2**: 95–98 (1983).

3.34. L. M. Biberman, *DAN SSSR* **59**: 659–661 (1948).

3.35. D. M. Pepper, *IEEE J. Quant. Electron.* **QE-14**: 971–977 (1978).

3.36. N. G. Preobrazhensky, in: *Proc. of 7th ESCAMPIG-84. Bari, Italy* (1984) : 182–185.

3.37. J. P. Limigne, J. P. Grandin, X. Husson, *et al.*, *J. Phys.* **44**, C7: 209–216 (1983).

3.38. N. G. Preobrazhensky, in: B. M. Smirnov, ed., *Novye napravleniya v optogalvanicheskoi spektroscopii* (New trends in optogalvanic spectroscopy) (Moscow: Energoatomizdat, 1987) (in Russian).

3.39. T. Berglind, L. Caspersson, *J. Phys.* **44**, C7: 329–334 (1983).

3.40. A. E. Bulyshev, N. V. Denisova, N. G. Preobrazhensky, *Optika i Spektroskopiya* **65**: 163–170 (1988).

3.41. W. L. Wiese, N. W. Smith, B. M. Glennon, *NBS Rept. NSRDS* **4** (1966).

3.42. L. Ph. Roesch, *Opt. Commun.* **44**:259–261 (1983).

3.43. L. M. Barkov, M. S. Zolotaryov, I. B. Khirplovich, *UFN* **132**: 409–442 (1980).

3.44. E. M. Anderson, V. A. Zilitis, *Optika i Spektr.* **16**: 382–389 (1964).

3.45. P. Chung, L. Talbot, K. Touryan, *Electric Probes in Stationary and Flowing Plasmas* (Heidelbereg: Springer Verlag, 1975).

3.46. F. Aeschbach, *Spectochim. Acta. B* **37**: 987–998 (1982).

3.47. R. J. Lovett, *Spectochim. Acta. B* **37**: 969–985 (1982).

3.48. I. J. M. M. Raaijmakers, P. W. J. M. Boumans, B. van der Sijde, D. C. A. Schram, *Spectrochim Acta. B* **38**: 697–706 (1984).

3.49. D. C. Schram, I. J. M. M. Raaijmakers, B. van der Sijde, *et al.*, *Spectrochim Acta. B* **38**: 1545–1557 (1984).

3.50. G. C. Turk, R. L. Watters, *Anal. Chem.* **57**: 1979–1983 (1985).

3.51. O. Axner, T. Berglind, *et al.*, *J. Phys.* **44**: 311–317 (1983).

3.52. N.B. Zorov, Yu. Ya. Kuzyakov, O. A. Novodvorsky, V. I. Chaplygin, in: B. M. Smirnov, ed., *Optogalvanichesky effekt v plamyonakh atmosfernogo davleniya* (Optogalvanic effect in atmospheric-pressure flames) (Moscow: Energoatomizdat, 1987): 131–163 (in Russian).

3.53. G. L. Moore, P. J. Humphries- Cuff, A. E. Watson, *Spectrochim. Acta. B* **39**: 915–929 (1984).

3.54. Kh. I. Zilbernshtein, M. A. Kartasheva, G. N. Moshkovich, and G. P. Startsev, in: *Optimizatsiya i issledovaniye vozmozhnostei metodov spektralnogo analiza vodnykh rastvorov s primeneniyem vysokochastotnogo induktivno-svyazannogo plazmennogo razryada* (Optimization and studies of the capabilities of spectral methods for the analysis of aqueous solutions with the use of inductively coupled plasma discharge) (Leningrad: Nauka, 1987): 94–114 (in Russian).

3.55. J. R. Branderberger, *Phys. Rev.* **36**: 76–80 (1987).

3.56. Yu. P. Raizer, M. N. Shneider, N.A. Yatsenko, *Radio-Frequency Capacitive Discharges* (NewYork: CRC Press, 1995).

3.57. B. I. Ilukhin, Yu. B. Udalov, I. V. Kochetov, V. N. Ochkin *et al.*, *Appl. Phys.* **62**: 113–127 (1996).

Chapter 4

Optogalvanic Effect in Conditions of Thermal Excitation of Atoms and Molecules

4.1 INTRODUCTORY REMARKS

This chapter considers the characteristic optogalvanic effect manifestations in flames and thermionic diodes. The optogalvanic effect in flames was observed for the first time in 1976 by Green and collaborators [4.1]. Today it is widely used at many laboratories, primarily because of the great applied capabilities of the optogalvanic detection technique in the field of atomic ionization analysis and also because of its potential in diagnosing burning processes [4.2–4.4].

The atomic-spectral method for determining elements on the basis of the optogalvanic effect in flames is noted for its express character and the relative ease of signal detection. One of its chief merits is its high sensitivity. The detection limit attained experimentally by the optogalvanic technique for traces of a wide variety of elements amounts to approximately 10^5–10^6 atoms/cm^3 [4.5] and in many cases greatly exceeds the limit reached by the other atomic-spectral analysis techniques [4.6]. The detection limit is mainly restricted by noise factors. The authors of [4.7, 4.8] have studied the effect of the easy-to-ionize impurities present in the flame, as well as the shape and arrangement of the electrodes, on the intensity of the noise component.

Another important aspect is the use of the optogalvanic spectroscopy of flames for diagnostic purposes. To illustrate, Lin and co-workers [4.9] used the optogalvanic effect to determine the temperature of flames by measuring the diffusion coefficient and mobility of ions. The figures obtained agreed well with the temperature measurement results obtained with other methods. The authors of [4.10, 4.11] determined the velocities of gases and flames. The idea was to focus laser radiation locally and then record the displacement of the laser-enhanced ionization region by the probe technique. The optogalvanic effect finds extensive application in the determination of various gas-kinetic parameters, for example, the diffusion coefficients and mobilities of ions. Such measurement methods were developed in [4.5, 4.12, 4.13]; by using the optogalvanic technique, the authors of [4.14] measured the recombination rate constant in a flame.

The mechanism of the optogalvanic effect in flames is usually associated with the fact that irradiation of the flame at a resonance transition frequency leads to an overpopulation of the excited state, a decrease in the ionization energy, and, as a consequence, an increase, according to Arrhenius, in the ionization rate constant. This may in turn cause a substantial reduction in the density of neutrals in the irradiation zone, which was clearly observed to occur in the experiment [4.21] for Na atoms.

Berglind and Casparsson [4.22] detected the optogalvanic effect in flames by probing the object of study with a microwave field. The attenuation η of the microwave field of frequency ω exhibits the following characteristic behavior:

$$\eta \sim (n/m)\left[v_\mathrm{t} / \left(\omega^2 + v_\mathrm{t}^2\right)\right],$$

where n and m are the charge density and mass, respectively and v_t is the frequency of collisions between the charged particle and its partners, the main contribution to absorption coming from electrons. Raising the field frequency $\omega > v_\mathrm{t}$ reduces absorption, and this places the upper limit on the field frequency. Technically, it is comparatively easy to measure relative microwave field power variations of around 10^{-3} at a total power of approximately 1 mW. Estimates show that the signal with such a probing causes no perceptible perturbations. This is one of the merits of the technique in comparison with the ordinary measurement schemes where the electrode potential amounts to some 10^3 V, which may noticeably perturb the distribution of the charged component. Another advantage of the technique is directly associated with its analytical applications and consists in the lowering of the element detection limit. The authors of [4.22] demonstrated that this advantage for Na was of the order of 10^3.

When interpreting optogalvanic measurements in flames, consideration should be given for the possibility of multiphoton ionization, for example, as is the case with the detection of the PO and NO radicals [4.23], as well as for that of stepwise photoionization, as suggested in interpreting the optogalvanic signals on alkali metal lines [4.3]. Reactions may also take place between the radiation-excited alkali metal atoms and electronegative molecules:

$$M^* + XY \to M^+ + XY^-.$$

It is important to note that the capabilities of the optogalvanic technique in the case of flames are not restricted to analytical applications and burning process diagnostics. They are much more extensive and will apparently be implemented in the near future, especially if the optogalvanic approach is used in conjunction with nonlinear spectroscopy and computer tomography techniques [4.15].

Until recently, the development of the optogalvanic technique in flames has been largely constrained, as in the case of electrodeless high-frequency discharge, by the lack of any acceptable model of the phenomenon. Only in the very recent past has progress been made along these lines.

4.2 MECHANISM GOVERNING THE OCCURRENCE OF OPTOGALVANIC SIGNALS IN FLAMES

As already noted, the optogalvanic effect is usually being investigated in flames at atmospheric pressure. In this case, use is made of commercially available burners employed in atomic absorption spectroscopy. The flame of interest is placed between electrode plates with a certain potential difference. The electrodes may be directly in the flame, or they may just touch it. The element under study in the form of a solution is atomized into the flame. The electrodes serve to detect an electric signal, for example, an electric current. When the flame is exposed to a resonance laser radiation, the strength of the current changes, and it is this change that is detected as the optogalvanic signal. The mechanism governing its occurrence is explained simply enough: the excitation of atoms by the laser radiation changes the ionization rate, hence the concentration of charged particles.

This in turn causes a change in the electrical (galvanic) characteristics of the flame.

If the flame is irradiated by a nanosecond-pulse laser, the shape of the optogalvanic signal in time is a double current spike. Figure 4.1*a* is a schematic diagram of the experimental setup used in [4.16], and Fig. 4.1*b* shows the optogalvanic signal obtained in this work, which can be considered to be typical A specific feature of this signal is that the first current spike is much greater in magnitude than the second, and the second spike is substantially delayed with respect to the first.

Several attempts were made to explain this signal shape. A series of works by the American group of researchers [4.1, 4.2, 4.16, 4.17, 4.21] were devoted to the development of the point-charge model. This model is based on the assumption that the charges formed at the irradiation point move independently under the effect of the external field. It is assumed that the electrodes are plane-parallel plates whose size is materially greater than the distance between them. In that case, there exists a direction selected along an axis normal to the electrode planes and the problem is

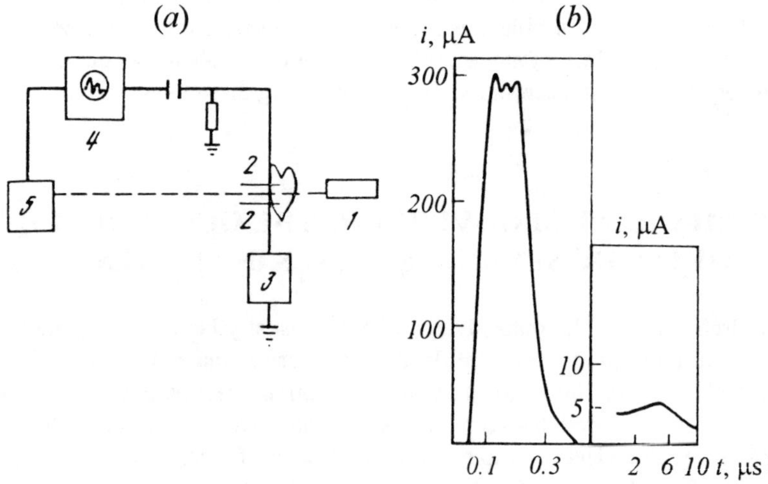

Figure 4.1. (*a*) Schematic diagram of the experimental setup and (*b*) the optogalvanic signal obtained. *1* – laser; *2* – electrode; *3* – power supply source; *4* – oscilloscope; *5* – photocell.

solved in the unidimensional approximation. At first the electric field strength distribution between the electrodes, E_{ext}, in the absence of laser irradiation is calculated on the basis of the Poisson equation. Next it is assumed that the charges $+q$ and $-q$ are formed at the instant $t = 0$ at the point with the coordinate $x = X_L$ (Fig. 4.2). In the given case, $q = Ne$, where N is the total number of charges (electrons or ions) formed at the irradiation point. The point-charge formation premise is justified in the case of crossed-beam laser irradiation scheme and two-step excitation, provided that the size of the region wherein the concentration of the charged particle rises is much smaller than the electrode separation. The external field E_{ext} causes the charges $+q$ and $-q$ to separate and move in the opposite directions, the positive charge moving towards the cathode and the negative one, towards the anode. The charges move with the velocities v_+ and v_- given by

$$v_d^{(i)} = \mu_i E_{\text{ext}} \text{ and } v_d = \mu_e E_{\text{ext}}, \tag{4.1}$$

where μ_i and $v_d^{(i)}$ are the ion mobility and drift velocity, respectively. These velocities are essentially different and are related together by the relation

$$\frac{v_d}{v_d^{(i)}} = \frac{\mu_e}{\mu_i} \cong 10^2 \text{ to } 10^3. \tag{4.2}$$

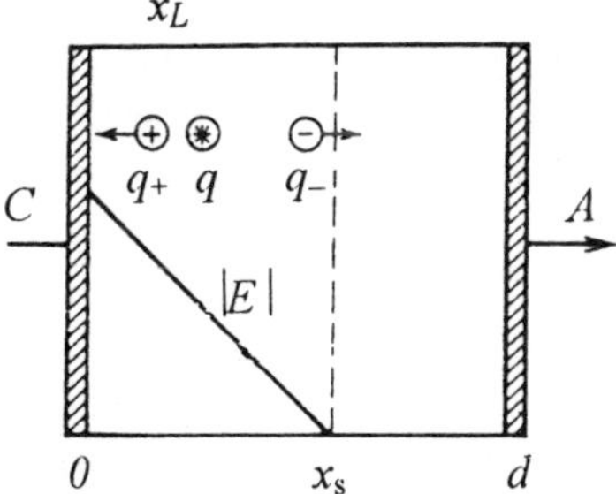

Figure 4.2. Idealized geometry of electric field distribution in a flame (unidimensional theory) and a point-charge model (C – cathode; A – anode).

This allows the motions of the positive and the negative charge to be considered independently of each other.

It is well known from the classical electrodynamics that any point charge located close to a conductive plate induces on the surface of the latter some charge of opposite sign. It has been assumed in [4.17] that a charge of q located at the point x induces point charges of q_0 and q_d on the electrode plates located at the points $x = 0$ (cathode) and $x = d$ (anode), these charges being expressed as

$$q_0 = -[(d - x)/d]q \text{ and } q_d = -(x/d)q. \tag{4.3}$$

Since the charge q (be it $+q$ or $-q$) is moving, a current is produced in the external circuit which is given, according to [4.17], by

$$i_q = -\frac{dq_0}{dt} = \frac{dq_d}{dt} = \frac{dx}{dt}\frac{q}{d} = \mu_{e,i} E_{ext} \frac{q_\pm}{d}. \tag{4.4}$$

It is assumed that the motion of the negative point charge gives rise to the first current spike i_e in the optogalvanic signal, while the motion of the positive charge is responsible for the second current spike i_i. The ratio between the current spike amplitudes is

$$i_e/i_i = \mu_e/\mu_i \cong 10^2 \text{ to } 10^3. \tag{4.5}$$

It should be noted that the results obtained in the above works have proven to be a good enough approximation, though they contain some disagreements with the experimental data. First, the results obtained in [4.1, 4.2, 4.16, 4.17, 4.21] only provide for linear relationship (4.4) between the optogalvanic signal and the external field, there being no provision for voltage saturation conditions. Secondly, the shapes of the current spikes i_e and i_i differ from the optogalvanic signal shape obtained experimentally (see Figs. 4.6 and 4.10).

The authors of [4.3, 4.18] conjectured that where the concentration of charged particles was as high as is typical of the irradiation region ($n \sim 10^{10} \text{ cm}^{-3}$), the charges of opposite signs could not separate because of the electric field produced between them, but formed a dipole instead. Qualitative considerations were also advanced as to the interpretation of the optogalvanic signal in that case. In the present section, we develop an approach which can be referred to as an electron-and-ion-cloud model. Within the framework of this model, use is made of the same geometry as that employed in the point-charge approximation (Fig. 4.2); the results obtained in this approximation are partially applicable here, too. The assumption that the charged particles formed occupy a certain space and the refinement of field distribution function $E_{ext}(x)$ has made it possible to eliminate the disagreements with experiment noted in [4.16, 4.17] and explain the optogalvanic signal voltage saturation effect. First, it is necessary to substantiate the very possibility of the decomposition of the plasmoid into an electron and an ion cloud and find out how independent their motions are relative to each other and what is their effect on the external field distribution.

4.3 ELECTRON AND ION CLOUD FORMATION PROCESSES

Assume that laser irradiation gives rise to a plasmoid at the focal spot, whose size is small compared to the electrode separation. Let us evaluate the possibility of its being separated into an electron and an ion cloud.

The plasma gets polarized under the effect of the external field E_{ext} (Fig. 4.3a). If the displacement of the charges relative to each other is small in comparison with the radius R_0 of the plasma cloud, the electric field set up between them can be estimated in the separate charged plate approximation:

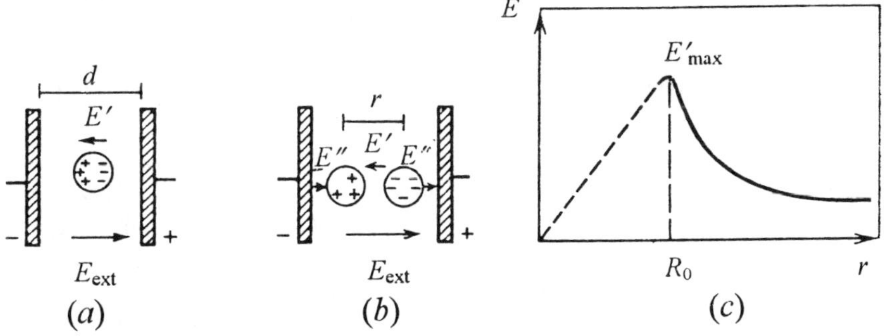

Figure 4.3. (a) and (b) Illustrating the formation of the electron and ion cloud and (c) the function $E'(r)$.

$$E' = 2\pi enr, \quad r \ll R_0, \tag{4.6}$$

where n is the concentration of charged particles in the plasma cloud and r is the distance that the charges are displaced.

If the displacement of the charged clouds is great, $r \geq R_0$ (Fig. 4.3b), the field E' produced between them is defined by the Coulomb law:

$$E' = ne4\pi R_0^3/3r^2, \quad r \geq R_0. \tag{4.7}$$

The notation here is the same as in Eq. (4.6).

Figure 4.3c qualitatively illustrates the behavior of the function $E'(r)$. The maximum strength of the field produced between the charges, which prevents them from being separated, may be estimated by the expression

$$E'_{\max} \cong ne4\pi R_0/3. \tag{4.8}$$

The condition for the plasma to become separated into an electron and an ion cloud may be formulated as follows:

$$E'_{\max} < E_{\text{ext}} \tag{4.9}$$

Otherwise, a dipole will be formed. Estimates yield $E'_{\max} = 1000$ V/cm at $n \sim 10^{10}$ cm^{-3}, $R_0 = 1$ mm. This is lower than the characteristic E_{ext} values known from the experimental data presented in [4.3, 4.16, 4.17]. The electron and ion clouds thus formed move in the external field independently of each other because their interaction through a distance of $r > R_0$ is negligibly weak. To illustrate, the field E' produced between two charged clouds with the same parameters ($n \sim 10^{10}$ cm^{-3}, $R_0 = 1$ mm) spaced a distance of $r \sim 1$ cm apart is estimated at around 3 V/cm. Consequently, the external field perturbation caused by the charged clouds is also negligible.

It should be noted that no account has been taken in estimates (4.6) and (4.7) of the field set up between the charged clouds and the electrode plates (E'' in Fig. 4.3b) and the screening of the electron and ion clouds by the charged particles in the surrounding plasma. Both these factors contribute to the separation of the plasma.

Thus, the estimates made for the characteristic experimental data presented in [4.1 – 4.3, 4.16, 4.17] show that the plasmoid under the effect of the external field gets separated into an electron and an ion cloud that move independently of each other, the perturbation of the field by these clouds being negligibly small.

4.4 INITIAL ASSUMPTIONS OF THE TWO-CLOUD MODEL

1. It is assumed that the laser irradiation of a flame placed between two electrodes gives rise to an enhanced ionization region – a plasmoid. According to the literature data [4.16, 4.17], the concentration n of charged particles in this region is of the order of 10^{10} cm^{-3}. It is also assumed that the size of the plasmoid is small compared to the interelectrode distance. The characteristic values indicated in [4.1–4.3, 4.16, 4.17] range from 1 to 2 cm, and the focal spot radius is $R_0 \cong 1$ mm.

Let us estimate the charged particle concentration in the laser-produced plasmoid on the basis of the experimental data presented in [4.16]. The current

$$i = n_e e E_{\text{ext}} \mu_e \pi R_0^2 \tag{4.10}$$

was measured in this work at 300 μA. According to [4.16, 4.17], the electron mobility μ_e in an air/acetylene flame at a temperature of 2500 K is of the order of 6×10^3 cm^2 s^{-1} at $E_{\text{ext}} \cong 2000$ V/cm and $R_0 \sim 1$ mm. So, out estimate yields $n_e \sim 10^{10}$ cm^{-3}.

2. In accordance with the considerations expounded in the preceding section, it is assumed that the laser-produced plasma breaks down under the effect of the external field into an electron and an ion cloud. These clouds move independently of each other towards the anode and the cathode, respectively. The rate at which the charged clouds move towards the electrodes is governed by their mobilities given by expressions (4.1). These expressions hold true when $\lambda_{e,i} \ll d$, where $\lambda_{e,i}$ is the electron or ion free path length and d is the electrode separation.

3. As with the point-charge model, we take the electrodes to be infinite planes. In that case, the problem becomes unidimensional. The edge effects are disregarded.

4. The motion of the charged particles is assumed to be due only to the external field, the role of their diffusion being negligible. To illustrate, for the ion current, we have

$$j_i = \mu_i E_{ext} n_i e + D_i e \frac{\partial n_i}{\partial x}, \qquad (4.11)$$

where D_i and n_i are the ion diffusion coefficient and concentration, respectively. The first term in Eq. (4.11) is associated with the drift of the ions and the second, with their diffusion. For estimation purposes, we put

$$E_{ext} \sim V/l \text{ and } \partial n_i/\partial x \sim n_i/d, \qquad (4.12)$$

where $l \sim 0.5.$ cm is the characteristic size and V is the potential difference between the electrodes. According to [4.17], the mobility of the Na$^+$ ions in an air/acetylene flame is $\mu_i = 25.8$ V$^{-1}\cdot$cm$^2\cdot$s^{-1} and their diffusion coefficient, $D_i \cong 5$ cm$^2\cdot$s^{-1} at a temperature of $T = 2500$ K. It follows from expressions (4.11) and (4.12) that the role of diffusion can be disregarded if $V >> D_i/\mu_i \cong 0.2$ V. A similar condition follows for the electron current as well. In the experiments described in [4.1–4.3, 4.16, 4.17], the potential difference V was 500 V and more. Consequently, one can take it that the motion of the electron and ion clouds is due to the field and the role of diffusion is indeed negligibly small.

5. Let us evaluate the role of bulk recombination in the "annihilation" of charged particles in the cloud, as compared with that of their drift-induced departure. The contribution from bulk recombination is negligible, provided that

$$\alpha n_i n_e^2 e << \frac{\partial j_{e,i}}{\partial x}, \qquad (4.13)$$

where α is the coefficient of bulk recombination in triple collisions. For estimation purposes, we put

$$\frac{\partial j_i}{\partial x} \sim n_i e \frac{V}{l^2}, \quad \frac{\partial j_e}{\partial x} \sim n_e e \frac{V}{l^2}. \qquad (4.14)$$

Substituting typical parameter values into expressions (4.13) and (4.14), we see that the role of bulk recombination can be disregarded at $V >> 10$ V. This condition is amply satisfied in experiments.

Our estimates show that the loss of electrons as a result of their attachment to oxygen molecules in three-body collisions in an air/acetylene flame is also small compared to their drift-induced departure.

4.5 ELECTRIC FIELD IN FLAMES

As demonstrated in Sect. 4.3 the effect of the charged clouds on the electric field distribution in the interelectrode space is negligibly small. The following approach is therefore possible to the solution of the problem: the distribution of the external field is first calculated in the absence of laser irradiation and then the motion of the charged clouds under the action of this field is analyzed.

The authors of [4.16, 4.17] calculated the electric field within the framework of the Tomson theory (see, for example, [4.19]). It was assumed in that case that the flame plasma in the region outside the cathode drop, $x > x_s$ (see. Fig. 4.2), was an ideal conductor. The effective anode was then placed at the point x_s and all the calculations were made for the domain $0 \leq x \leq x_s$.

The following field distribution was obtained as a result:

$$E_{\text{ext}} = \delta(x_s - x), \qquad x \leq x_s, \qquad (4.15)$$

$$E_{\text{ext}} = 0, \qquad x > x_s. \qquad (4.16)$$

According to the calculations made in [4.17],

$$\delta = (r_c e/\mu_i)^{1/2}, \qquad (4.17)$$

$$x_s = (2V/\delta)^{1/2}, \qquad (4.18)$$

where r_c is the ionization factor in the flame and V is the difference in potential between the electrodes. Figure 4.4 presents experimental data on the measurement of the electric field distribution in the interelectrode space. In accordance with these data and also the measurement results obtained in [4.3, 4.16], it seems more correct to choose the external field distribution function in the following form:

$$E_{\text{ext}} = \delta(x_s - x), \qquad x \leq x_0, \qquad (4.19)$$

$$E_{\text{ext}} = E_0, \qquad x > x_0, \qquad (4.20)$$

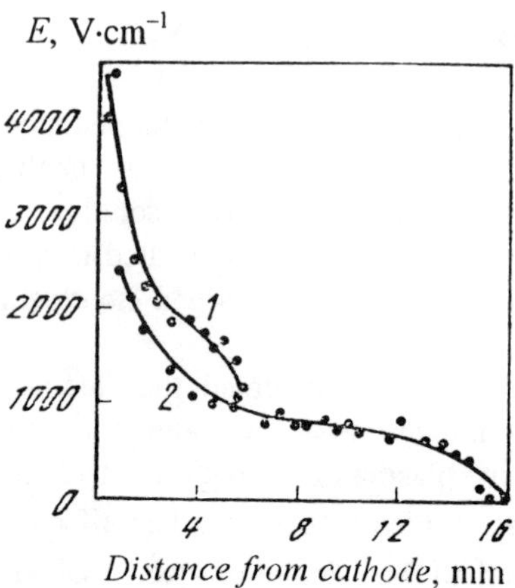

Figure 4.4. Electric field in a flame. Experimental data borrowed from [4.16]. $V =$ 1500 V, electrode separation d: $1 - 1$ cm; $2 - 2$ cm.

where δ and x_s are defined in accordance with expressions (4.17) and (4.18) and E_0 and x_0 can, in principle, be calculated within the framework of the Tomson approximation, but in the given case they are found from experimental data.

Figure 4.5 shows a model distribution of the electric field E_{ext} in the interelectrode space, calculated in accordance with formulas (4.19) and (4.20). This distribution is used in this book to calculate the motion of charged clouds. It will be demonstrated elsewhere that the correction of E_{ext} (the introduction of E_0) enables one better to match the theoretical optogalvanic signal to the experimental data in hand.

In the text to follow, δ and x_s are taken to be defined by Eqs. (4.17) and (4.18), respectively, and E_0, by Eq. (4.20).

4.6 FIRST OPTOGALVANIC SIGNAL PULSE

In accordance with the model described above, one should take it that the first optogalvanic signal pulse is due to the motion of the electron cloud. It includes two phases – the current associated with the motion of the electron cloud in the interelectrode space and the current developing

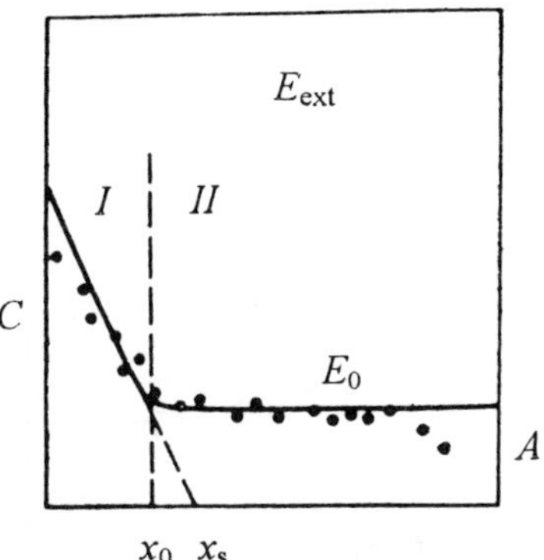

Figure 4.5. Electric field in a flame. Theoretical interpolation of experimental data. C – cathode; A – anode.

directly as a result of this cloud's arrival at the anode. It is necessary to reveal the contribution from each phase to the optogalvanic signal.

The authors of [4.16, 4.17] calculated the first phase of the optogalvanic signal within the framework of the point-charge model (see Eqs. (4.3) and (4.4)). By substituting into Eq. (4.4) the expression for E_ext from (4.15) and putting $q_- = Ne$, (N being the total number of electrons produced by radiation), the following expression was obtained in [4.16] for the electronic current pulse:

$$i = Ne(x_s - x_L)\exp(-t/\tau_-)/x_s\tau_-, \qquad t \geq 0,$$

$$\tau_- = (\mu_e\delta)^{-1},$$

(4.21)

where x_L is the coordinate of the irradiation point. Figure 4.6 compares these results with experimental data. The theoretical signal can be seen to be much narrower than its experimental counterpart. This can be explained by the fact that the approximation of the external field distribution in [4.17] was not correct enough. We will assume that the behavior of the filed conforms with expressions (4.19) and (4.20). In that case, the motion of the electron cloud in this field in the region I to left of x_0 (see Fig. 4.5) gives rise to a current corresponding to Eq. (4.21). The width of this signal portion in time is found subject to the condition

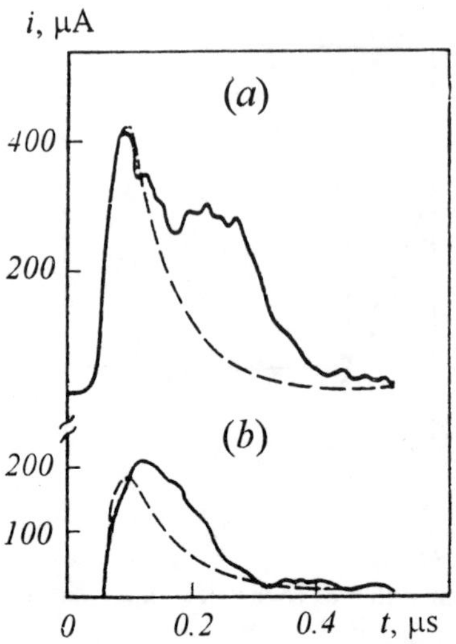

Figure 4.6. Electronic-pulse portion of the optogalvanic signal. The distance from the cathode to the laser irradiation point, x_L: (a) 1 mm; (b) 9 mm. The solid curves – experiment, and the dashed ones, calculation [4.17].

$$\frac{dx}{dt} = \mu_e \delta(x_s - x)$$

to be

$$\Delta t_1 = \frac{1}{\delta \mu_e} \ln\left(\frac{x_s - x_L}{x_s - x_0}\right). \tag{4.22}$$

Next the motion of the cloud takes place in the field region II to the right of the point x_0 (see Fig. 4.5), where $E = E_0$. The density of the current induced by the motion of the charged cloud is

$$j = n_e e \mu_e E_0. \tag{4.23}$$

Expression (4.23) is graphically represented by a plateau whose width on the time scale is given by

$$\Delta t_2 = (d - x_0)/\mu_e E_0. \tag{4.24}$$

If the laser radiation is focused into the region II to the right of the point x_0 (Fig. 4.5), the optogalvanic signal fully corresponds to the plateau of the curve defined by Eq. (4.23) The height of the plateau is independent of time, and its width is directly proportional to the distance between the irradiation point x_L and the anode. These results are confirmed by the experimental data obtained in [4.16] (see Fig. 4.6b).

Theoretically, the current i_e should develop the instant the electron cloud starts moving. It was supposed in [4.3, 4.18] that the delay of the signal was associated with the relaxation of the plasma and was of the order of $\tau_{\text{rel}} \sim 1/4\pi\sigma \sim 10^{-8}$ s, where σ is the conductivity of the flame plasma. This might also be due to the long time constant of the amplifier used.

The arrival of the electron cloud directly at the anode also contributes to the optogalvanic signal. This process is mathematically described by means of the charged particle balance equation. If diffusion and recombination are disregarded, this equation is written as

$$\frac{\partial n_e}{\partial t} = \frac{\partial F_e}{\partial x},\tag{4.25}$$

where the electron flux is given by

$$F_e = \mu_e |E_{\text{ext}}| n_e.$$

Let us define the current density:

$$j = eF_e = e\mu_e |E_{\text{ext}}| n_e.$$

In that case, Eq. (4.25) may be written for the current density as follows:

$$\left(\mu_e |E_{\text{ext}}|\right)^{-1} \frac{\partial j}{\partial t} + \frac{\partial j}{\partial x} = 0, \quad 0 \le x \le d.\tag{4.26}$$

Consider this equation for the case of cloud motion in the field $E = E_0$. Let the region corresponding to the laser spot have the boundaries $[x_1, x_2]$. We assume that at the instant $t = 0$ there is produced in this region the initial electron flux

$$j(t=0) = \begin{cases} 0, & x < x_1, \\ j_0 = e\tau_L r_L \mu_e E_0, & x_1 \le x \le x_2, \\ 0, & x > x_2, \end{cases} \qquad (4.27)$$

where τ_L is the laser pulse duration and r_L is the rate of bulk recombination due to laser radiation. The solution of Eq. (4.26) at the anode, i. e., for $x = d$, subject to initial conditions (4.27) is as follows:

$$j(x=d) = \begin{cases} 0, & t < (d - x_2)/\mu_e E_0, \\ j_0, & (d - x_2)/\mu_e E_0 \le t \le (d - x_1)/\mu_e E_0, \\ 0, & t > (d - x_1)/\mu_e E_0. \end{cases} \qquad (4.28)$$

The duration of the optogalvanic signal associated with the arrival of the electron cloud at the anode is

$$\Delta t_3 = (x_2 - x_1)/\mu_e E_0. \qquad (4.29)$$

Thus, the overall electronic current pulse can be represented in the shape shown in Fig. 4.7. Here I is the current variation region defined by expression (4.21), with Δt_1 corresponding to Eq. (4.22), II is the plateau defined by Eq. (4.23), whose width is Δt_2 given by Eq. (4.24), and III is the region corresponding to the arrival of the electron cloud at the anode, defined by Eq. (4.28), its width Δt_3 being defined by Eq. (4.29). Let us estimate Δt_1, Δt_2, and Δt_3 for the characteristic parameters corresponding to the experimental data of [4.16]: $x_2 - x_1 = 1$ mm, $x_s = 1$ cm, $d = 2$ cm, $\Delta t_1 \sim 3 \times 10^{-8}$ s, $\Delta t_2 \sim 2 \times 10^{-7}$ s, and $\Delta t_3 \sim 2 \times 10^{-8}$ s.

On the whole, these estimates agree with the results obtained experimentally in [4.16]. One should also note the following circumstances. As the charged cloud moves in the external field, it spreads on account of repulsive electrostatic forces. The expansion of the cloud boundaries can be approximately estimated by analogy with the streamer discharge theory [4.20] as follows:

$$dR/dt = \mu_e E = \mu_e Ne/R^2, \qquad (4.30)$$

where E is the field strength on the cloud surface. Solving this equation yields

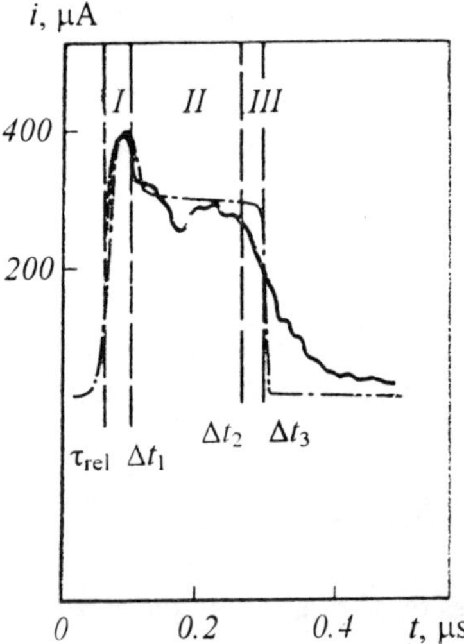

Figure 4.7. Electronic-pulse portion of the optogalvanic signal. The solid curve –
experiment [4.16], and the dashed curve represents the present calculation.

$$R = (3\mu_e Net + R_0^{\,3})^{1/3}. \qquad (4.31)$$

The spreading rate of the cloud is

$$dR/dt = \mu_e Ne(3\mu_e Net + R_0^{\,3})^{2/3}. \qquad (4.32)$$

Estimates show that for real experimental conditions ($n_e \sim 10^{10}$ cm^{-3}),
the radius of the electron cloud increases two or three times by the instant
it reaches the anode. This increases the Δt_3 value calculated earlier the
same number of times, and reduces accordingly the Δt_2 value, but the shape
of the optogalvanic signal in that case suffers no substantial changes.

4.7 SECOND OPTOGALVANIC SIGNAL PULSE

The second optogalvanic signal pulse is associated with the motion of the
ion cloud. Inasmuch as the motion velocities of the electron and ion clouds
differ by two to three orders of magnitude and their interaction is
negligibly small, their motions can be analyzed on different time intervals

and independently of each other. The ion cloud moves in the field defined by Eq. (4.19). As in the case of electronic pulse, we consider two phases of the optogalvanic signal.

The following expression was obtained in [4.17] for the ion current due to the motion of the ion cloud in the interelectrode space:

$$i_i = \begin{cases} Ne(x_s - x_L)\exp\!\left[(t/\tau_i)/x_s\tau_i\right], & t_{arr} > t > 0, \\ 0, & t > t_{arr}, \end{cases} \qquad (4.33)$$

where $\tau_i = (\mu_i \delta)^{-1}$ and t_{arr} is the time of the ion cloud arrival at the cathode, given by

$$t_{arr} = \tau_i \ln\left[x_s/(x_s - x_L)\right]. \qquad (4.34)$$

Figure 4.8. compares these results with the experimental data presented in [4.16].

The expansion of the ion cloud can be allowed for in accordance with Eqs. (4.30)–(4.32), with μ_e being replaced by μ_i. Estimates give 1.5- to 2-fold increase in the ion cloud radius. Figure 4.8 also presents the results of calculating the optogalvanic signal with correction made to the time of

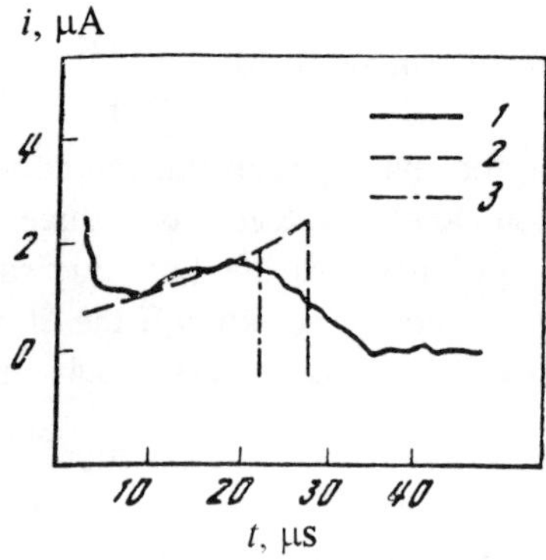

Figure 4.8. Ionic-pulse portion of the optogalvanic signal. *1* – experimental data [4.16]; *2* – calculation results [4.17]; *3* – correction to the arrival time of the ion cloud at the cathode, made to allow for its expansion (calculated on the basis of experimental parameters [4.16]).

arrival of the electron cloud at the cathode in order to allow for its expansion. The authors of [4.17] believed that as soon as the ion cloud arrived at the cathode, the optogalvanic signal current went to zero.

Next we consider the contribution to the optogalvanic signal from the current associated with the arrival of the ion cloud directly at the cathode. By analogy with Eqs. (4.25) and (4.26), the equation for the ion current has the form

$$(\mu_i\delta)^{-1}(x_s - x)^{-1}\partial j_i/\partial t - \partial j_i/\partial x = 0. \tag{4.35}$$

By analogy with expressions (4.27), the initial conditions are as follows:

$$j_i(t=0) = \begin{cases} 0, & x < x_1, \\ j_0, & x_1 \le x \le x_2, \\ 0, & x > x_2, \end{cases} \tag{4.36}$$

$$j_0 = e r_L \tau_L \delta(x_s - x)\mu_+. \tag{4.37}$$

Let us write t_1 for the time of arrival of the ion cloud at the cathode and t_2 for the signal termination time. Based on the condition that

$$dx/dt = \mu_i\delta(x_s - x), \tag{4.38}$$

we find

$$t_1 = (\mu_i\delta)^{-1} \ln [x_s/(x_s - x_1)], \tag{4.39}$$
$$t_2 = (\mu_i\delta)^{-1} \ln [x_s/(x_s - x_2)]. \tag{4.40}$$

We will seek the solution of Eq. (4.33) in the form

$$j_i = j_i(\xi), \tag{4.41}$$

where ξ is defined as follows:

$$\xi = -\ln (x_s - x) + \mu_i\delta t \tag{4.42}$$

Based on initial conditions (4.36), we find $j_i(\xi)$:

$$j_i(t=0) = \begin{cases} 0, & x < x_1, \\ er_L\tau_L\delta(x_s - x), & x_1 \leq x \leq x_2, \\ 0, & x > x_2. \end{cases} \qquad (4.43)$$

In that case,

$$j_i(x, t) = er_L\tau_L(x_s - x)\delta \exp(-\mu_i\delta t), \qquad 0 \leq x \leq d. \qquad (4.44)$$

The ion current at the cathode ($x = 0$) is

$$j_i(x=0) = \begin{cases} 0, & t < t_1, \\ er_L\tau_L x_s \exp(-\mu_i\delta t), & t_1 \leq t \leq t_2, \\ 0, & t > t_2. \end{cases} \qquad (4.45)$$

The values of t_1 and t_2 should be corrected to allow for the expansion of the ion cloud. Figure 4.9 shows the second optogalvanic signal pulse associated with the arrival of the ion cloud directly at the cathode. Region *I* corresponds to the first phase – the current induced by the motion of the ion cloud in the interelectrode space and region *II*, to the arrival of the ion cloud directly at the cathode. A good agreement with experimental data exists.

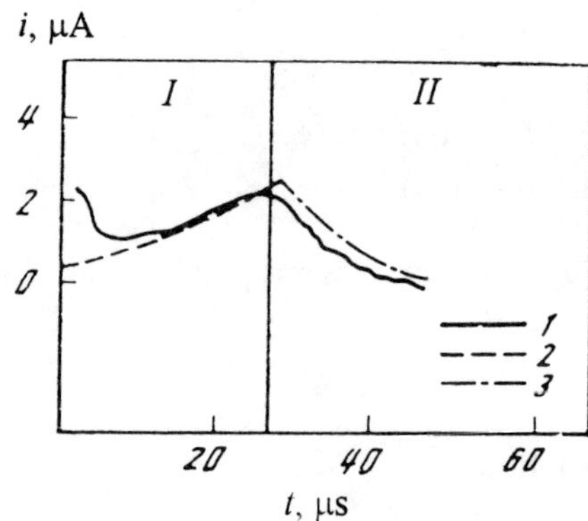

Figure 4.9. Ionic-pulse portion of the optogalvanic signal. *1*, *2*, and *3* – the same as in Fig. 4.8.

4.8 OPTOGALVANIC SIGNAL VOLTAGE SATURATION

Figure 4.10 presents the relationship between the amplitude of the electronic optogalvanic signal and the voltage across the electrodes, borrowed from [4.7]. As one can see, the signal amplitude first grows linearly with the increasing voltage. This result is contained in Eqs. (4.4) and (4.27). However, as the voltage grows higher, the signal is observed to suffer saturation. This effect can be explained as follows. If the laser pulse duration τ_L is short in comparison with the time t_f it takes for the charged particles to fly a distance of the order of the laser spot size $2R_0 = (x_2 - x_1)$,

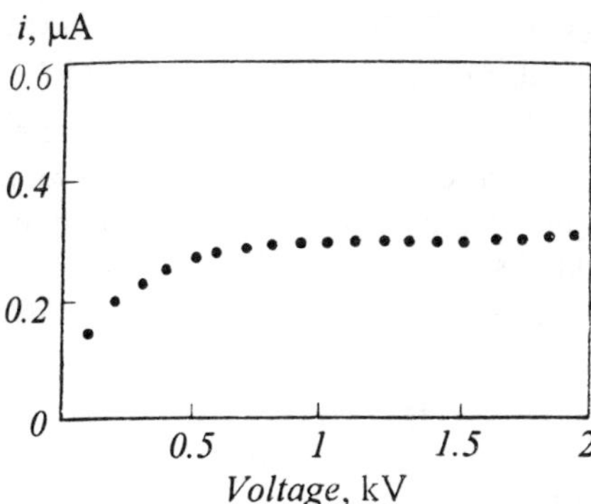

Figure 4.10. Optogalvanic signal voltage saturation.

the flux of the charged particles produced obeys Eq. (4.27) and expresses the initial conditions of the problem. If the condition

$$\tau_L \geq t_f \qquad (4.46)$$

is satisfied, the flux of the charged particles produced in the irradiation region will be

$$j_e = rt_f^- e\mu_e E, \qquad j_i = r_L t_f^+ e\mu_i E, \qquad (4.47)$$

where t_f^- is the time the electrons take to cover the distance $(x_2 - x_1)$ and t_f^+, the same for the ions.

Obviously

$$t_f^- = (x_2 - x_1)/\mu_e E \text{ and } t_f^+ = (x_2 - x_1)/\mu_i E \tag{4.48}$$

Substituting expressions (4.48) into (4.47), we find that the current developing in this case is independent of the applied voltage:

$$j_e = r_L e(x_2 - x_1), \qquad j_i = r_L e(x_2 - x_1) \tag{4.49}$$

Thus, condition (4.46) can be considered the condition for the saturation of the signal. Obviously, the electronic pulse is saturated at much lower applied voltages than its ionic counterpart. It was only the electronic signal that was experimentally observed in [4.5] to undergo saturation at a voltage of $V_{sat} = 600$ V, and so in the text to follow we will only consider this signal. Estimates by formula (4.46) for $\tau_L \sim 10^{-8}$ s and $(x_2 - x_1) = 1$ mm yield $V_{sat} \geq 500$–1000 V, which agrees well with the experimental results. Approaching the saturation problem from the standpoint of the gas discharge theory [4.19], one can consider the quasistationary conditions under which the following saturation current develops in the region $(x_2 - x_1)$:

$$j_e = e r_L(x - x_1), \quad x \leq x \leq x_2. \tag{4.50}$$

We assume for simplicity that irradiation occurs in the field region II where $E = E_0$. The induced current defined by expression (4.23) is also saturated. Indeed, the electron density in that case is

$$n_e = r_L t_f^- = r_L(x_2 - x_1)/\mu_e E_0, \tag{4.51}$$

and the current density

$$j_e = n_e e \mu_e E_0 = e r_L(x_2 - x_1). \tag{4.52}$$

We write t_1' for the time of arrival of the electron cloud at the anode and t_2' for the optogalvanic signal termination time:

$$t_1' = (d - x_2)/\mu_e E_0, \qquad t_2' = \tau_L + (d - x_1)/\mu_e E_0. \tag{4.53}$$

And the density of the current associated with the arrival of the electron cloud at the anode will be

$$
j_e(x=d) = \begin{cases} er_L(x_2 - x_1), & t_1' \le t \le \tau_L + t_1', \\ er_L(x_2 - x_1) - \mu_e E_0\left(t - \tau_L - t_1'\right), & \tau_L + t_1' \le t \le t_2', \\ 0, & t > t_2'. \end{cases} \quad (4.54)
$$

Figure 4.11 shows the shape of the optogalvanic signal under saturation conditions. Here I stands for the induced current region corresponding to Eq. (4.52), and II and III denote the regions associated with the arrival of the electron clod at the anode, where the current density is defined by expressions (4.54). Such a signal shape was observed in the experiment [4.16] at high voltage values ($V \sim 1500$ V) when the laser beam was focused into a region well off the cathode: $x_L \approx 9$ mm (see Fig. 4.6b). Thus, the physical model being considered for the optogalvanic effect in flames, based on the assumption that the flame plasma gets separated into an electron and an ion cloud, adequately describes the characteristics of the optogalvanic signal observed experimentally.

As for the earlier point charge model, it may prove acceptable for approximate description purposes in certain situations (the first phase of the current pulse) (see [4.16]), but it cannot help to correctly calculate the

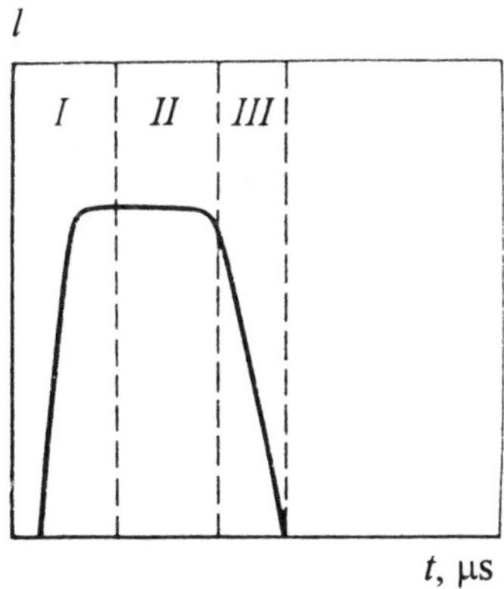

Figure 4.11. Optogalvanic signal shape under voltage saturation conditions.

second signal phase, nor can it explain the voltage saturation of the signal or provide for any better agreement between quantitative calculation results and experimental data even for the first signal phase.

4.9 THERMIONIC DIODES

This device (also known as the thermionic transducer or space-charge-limited diode) was used as an ion detector as far back as 1923 [4.25, 4.26]. A thermionic diode in its simplest form comprises a cathode (a hot wire) and an anode a few millimeters distant from it, usually in the form of a cylinder enveloping the cathode wire (Fig. 4.12). The emission of electrons from the cathode produces a negative space charge region. The electrons may overcome this potential barrier and cause a current to flow in the diode circuit without any voltage applied across the electrodes, provided the latter are made of materials differing in work function. If an ion finds

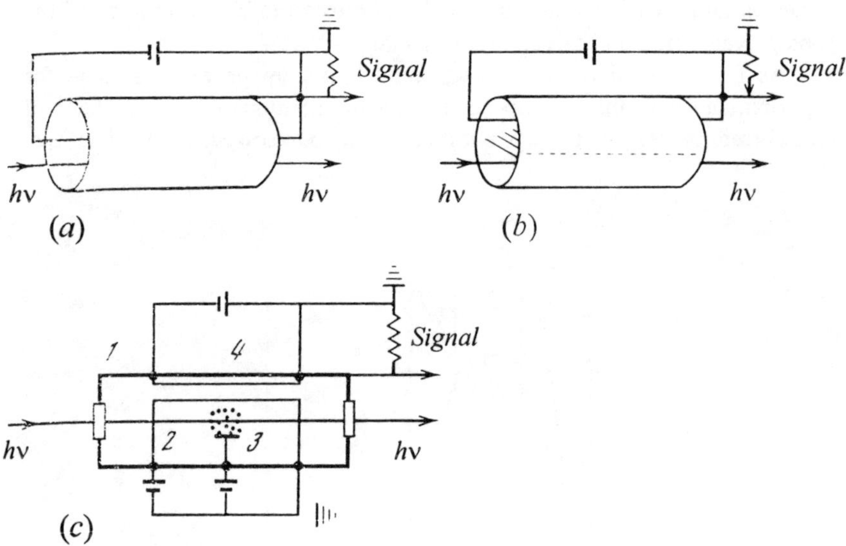

Figure 4.12. Thermionic diodes: (*a*) single diode; (*b*) single diode with a screen grid; (*c*) double-diode. *1* – tube; *2* and *3* – cathode and anode of the exciting diode, respectively; *4* – detector diode.

its way into the space charge region, it lowers the potential barrier while moving toward the cathode, which becomes manifest as an increase in the current through the diode. The diode is an ion trap: ions cannot leave the diode through its end faces because the lower temperature at the ends of the hot cathode wire reduces the space charge density there. The small electron-ion recombination cross sections provide for a prolonged residence of ions in the volume of the diode. As a result, the device is characterized by a high sensitivity, and as shown by the experiments conducted in [4.27], even so low an ion influx as 10 ions per second causes a diode current change exceeding the noise level.

The thermionic diode was used for the first time for optogalvanic spectroscopy purposes in [4.28]. Some aspects of the spectroscopy of highly excited states using such devices were expounded in [4.29]. For highly excited Rydberg states, the space charge field may cause their Stark perturbation. To avoid this effect, the diode used in [4.30] was separated by a screen grid (Fig. 4.12b). Atoms were excited in the bottom diode region screened from the field and diffused into the space charge region. However, when experimenting with alkaline earth metals, when high temperatures are required to attain the necessary particle density, practically all the elements of such a diode structure prove emitting. This results in the levels with $n \geq 100$ experiencing a strong influence from the field produced by the charges. To cover the region of levels with $n \geq 100$, the authors of [4.31] proposed an annular diode using not one but several hot wires arranged symmetrically with respect to the diode axis. By virtue of symmetry, the field at the axis is virtually nonexistent. Niemax and Lorenzen [4.32] proposed a thermionic double-diode (Fig. 4.12c). The idea is to preliminarily excite the atoms of interest in a discharge and then study their absorption spectra. The diode structure is enclosed in a heated steel tube 1. The exciting diode consists of a hot cathode 2 and an anode 3 – a metal plate 3–4 mm distant from the cathode. A d. c. voltage (a few volts at some 10^{-1} Torr of the metal vapor in an inert buffer gas at a pressure around 10^{-1} Torr) produces a weak glow discharge illuminated by a laser. The laser-excited atoms become ionized either in the discharge or in the space charge region of the second, thermionic diode and are detected there. The use of the double-diode provides for a sensitivity 10^2–10^3 times better than in the case of optogalvanic signal detection in a glow discharge. To illustrate, comparison between the line profiles of the $5s5p\,^1P_1 \rightarrow 5s7s\,^1S_0$ transition in strontium, obtained by thermionic diode detection and by direct optogalvanic signal detection in the exciting diode under optimum discharge conditions, has shown that the gain in the signal/noise ratio in the former case amounts to around 400 [4.32].

The thermionic diode technique to be used in conjunction with optogalvanic detection for excited state laser spectroscopy purposes has attracted considerable attention and continues to be improved. For example, Hoffnagle and Schlumpf [4.33] have suggested a diode design combining the ideas of the electrostatically shielded diode and the double-diode. In that case, however, the atoms are excited not in an auxiliary discharge, but directly by the electrons emitted by the hot cathode.

REFERENCES

4.1. R. B. Green, R. A. Keller, P. K. Schenk *et al.*, *J. Amer. Chem. Soc.* **98**: 8517–8518 (1976).

4.2. G. C. Turk, J. R. DeVoe, J. C. Travis, *Anal. Chem.* **54**: 643 – 645 (1982).

4.3. N. B. Zorov, Yu. Ya. Kuzyakov, O.A. Novodvorsky, V. I. Chaplygin, in: B.M. Smirnov, ed., *Khimiya Plasmy: Sb. statei* (Plasma chemistry: Collected papers) (Moscow: Energoatomizdat, 1987): 131 – 163 (in Russian).

4.4. V.N. Ochkin, N. G. Preobrazhensky, . N. Sobolev, N. Ya. Shaparev, *UFN* **148**, No. 3: 473–507 (1986).

4.5. N. B. Zorov, Yu. Ya. Kuzyakov, O. I. Matveyev, V. I. Chaplygin, *Zhurn. Analit. Khimii* **35**: 1701–1707 (1980).

4.6. J. C. Travis, G. C. Turk, R. B. Green, *Anal. Chem.* **54**: 1006–1018 (1982).

4.7. G. C. Turk, *Anal. Chem.* **53**: 1187–1190 (1981).

4.8. O. I. Matveyev, *Optika I Spektroskopiya* **63**, No. 5: 1171 – 1174 (1987).

4.9. K. C. Lin, P. M. Hunt, S. R. Crouch, *Chem. Phys. Lett.* **90**: 111–116 (1982).

4.10. P. K. Schenk, J. C. Travis, G. C. Turk, T. C. O'Haver, *Appl. Spectr.* **36**: 168–171 (1982).

4.11. Yu. Ya. Kuzyakov, O. I. Matveyev, O. A. Novodvorsky, *Zhurn. Prikl. Spektroskopii* **40**, No. 1: 145–148 (1984).

4.12. W. G. Mallard, K. C. Smyth, *Comb. Flame* **44**: 61–70 (1982).

4.13. O. A. Novodvorsky, A. B. Ilyukhin, Yu. Ya. Kuzyakov, *Zhurn. Prikl. Spektroskopii* **47**, No. 3: 385–389 (1987).

4.14. A. M. Udartsev, V. Kim, G. K. Iordanidi, S. M. Mashakova, and G. I. Ksandopulo, *Zhurn. Prikl. Spektroskopii* **46**, No. 1: 38 – 41 (1987).

4.15. N. G. Preobrazhensky, in: B. M. Smirnov, ed., *Khimiya plazmy: Sb. statei* (Plasma chemistry: Collected papers) (Moscow: Energoatomizdat, 1987): 114–131 (in Russian).

4.16. G. J. Havrilla, P. K. Schenk, J.C. Travis, G.C. Turk, *Anal. Chem.* **56**: 186–193 (1984).

4.17. J. C. Travis, G. C. Turk, J. R. De Voe, *et al.*, *Progr. Analyt. Atom. Spectrosc.* **7**: 199–241 (1984).

4.18. O. A. Novodvorsky, N. B. Zorov, Yu. Ya. Kuzyakov, *Vesti MGU. Ser. 2, Khimiya* **26**, No. 2: 221–222 (1985).

4.19. V. L. Granovsky, *Elektrichesky tok v gaze* (Electric current in gases) (Moscow: Nauka, 1971) (in Russian).

4.20. Yu. P. Raizer, *Fizika gazovogo razryada* (Gas discharge physics) (Moscow: Nauka, 1987) (in Russian).

4.21. P. K. Schenk, J.C. Travis, G. C. Turk, *J. de Phys.* **44**, C7: 75 – 84 (1983).

4.22. T. Berglind, L. Casparsson, *J. de Phys.* **44**, C7: 329–334 (1983).

4.23. K. C. Smyth, W. G. Mallard, *J. Chem. Phys.* **77**: 1779–1787 (1982).

4.24. I. M. Beterov, A. V. Eletsky, B.M. Smirnov, *UFN* **155**: 265 – 298 (1988).

4.25. G. Hertz, *Z. Phys.* **18**: 307–316 (1923).

4.26. K. H. Kington, *Phys. Rev.* **21**: 408–418 (1923).

4.27. D. Popescu, I. I. Popescu, J. Richter, *Z. Phys.* **226**: 16—174 (1969).

4.28. P. D. Foote, F. I. Mohler, *Phys. Rev.* **92**: 896–898 (1953).

4.29. B. M. Smirnov, *Vozbuzhdyonnye atomy* (Excited atoms) (Moscow: Energoatomizdat, 1982) (in Russian).

4.30. K. C. Harvey, *Rev. Sci. Instr.* **52**: 204–206 (1981).

4.31. R. Beigang, A. Timmerman, *J. de Phys.* **44**, C7: 137–148 (1983).

4.32. K. Niemax, C.-J. Lorenzen, *Opt. Comm.* **44**: 165–169 (1983).

4.33. J. Hoffnagle, N. Schlumpf, *Opt. Comm.* **55**: 325–328 (1985).

Chapter 5

High-Spectral-Resolution Optogalvanic Detection

5.1 INTRODUCTORY REMARKS

Recently, there is, a trend to combine the optogalvanic detection techniques with the linear and nonlinear laser spectroscopy, level-crossing spectroscopy, double-resonance spectroscopy, and other spectroscopic techniques that have gained wide recognition on a purely optical basis. The product of this synthesis is a wide variety of highly sensitive, high-precision methods for studying atoms, ions, and molecules (free radicals included), as well as their interaction with one another and with external fields, which are comparatively easy to implement.

By using the multiple-step laser pumping principles, one can achieve complete ionization of the atoms and molecules excited in a resonance fashion, with the participation of collisional processes being completely excluded. In other words, one can succeed in transforming in a purely radiative way a neutral particle of a given species into a pair of charged particles – an ion and an electron – which can then de detected with ease. This approach is at the root of laser resonance photoionization spectroscopy [5.1] which possesses a record-high sensitivity and makes it possible to detect individual atoms, molecules, or ions. At the same time, one should not forget that such techniques are rather costly and complex, and are subject to a number of restrictions as to the range of objects studied, and are rather slow.

The analysis of the capabilities of both the sub-Doppler optogalvanic spectroscopy technique based on a low-pressure gas discharge and the as yet rather exotic multistep photoionization technique enables one to reveal the strong and weak sides of either approach.

It should be emphasized that this technique is by no means any ousting of the common nonlinear spectroscopy, degenerate-state interference, and other traditional techniques [2.5] that have already been universally accepted and proved their worth time and again. The optogalvanic spectroscopy of narrow nonlinear resonances, with its specific features and preferable fields of application, proves an extremely useful supplement to those methods wherein physically similar resonances are recorded in a purely optical way.

The time resolution of the present-day laser spectroscopy techniques may also be very high, reaching a few tens of nanoseconds for atoms, and for molecules, even a few picoseconds [5.1–5.3]. It is entirely possible to produce conditions under which the homogeneous broadening of an absorption spectral line in a gas or in a low-pressure gas-discharge plasma noticeably exceeds the broadening due to the relaxation of the excited levels. As a result, such a time resolution is attained as depends on the inverse homogeneous half-width Γ_0^{-1}, though if the influence of the Doppler effect is eliminated (see below), the half-width Γ_0 itself characterizes the spectral resolution of the optogalvanic technique. It is important that when one works with picosecond pulses [5.3], one can investigate, for example, the excited state dynamics in molecules under "collisionless" conditions even at not very low gas pressures.

And finally a few remarks on the spatial resolution of the optogalvanic signal: in the majority of experiments, the optogalvanic effect is assumed to be integral from the standpoint of accumulation of the excited atoms of molecules all over the length of the exciting laser beam within the object of interest. Accordingly, the integral along the beam becomes the optogalvanic signal detected. However, the localization of the signal, that is, its reference to some region of the gas (or plasma) medium, is, in fact, inevitable in solving diagnostic problems where the object is essentially inhomogeneous. The methods developed for direct optogalvanic localization usually combine well with the nonlinear spectroscopy and double-resonance techniques considered below, as well as with the highly sensitive photoionization spectroscopy technique [5.1]. Another approach to the problem is associated with the use of the computer tomography principles [5.4–5.6]. The latter technique is advantageous because it is not accompanied with any artificial narrowing of the laser-plasma interaction region and thus entails no loss of signal power. Moreover, given certain generalizations (the so-called chronotomography), it can, in principle, provide for the integral solution of the optogalvanic problem, with both high spatial resolution and high time resolution at the same time.

5.2 LASER-INDUCED CURRENTS IN RAREFIED GASES AND PLASMAS

We have satisfied ourselves time and again that the interpretation of the optogalvanic spectra detected by means of some or other type of gas discharge or flame is usually in no way conspicuous for reliability, and as far as quantitative calculations are concerned, requires much effort, because most of the known theoretical models are both imperfect and involved. In this connection, it would be of interest to realize such conditions as would be conducive to the development of the light-induced current in a sufficiently rarefied (Knudsen's) gas or rarefied plasma, for in that case it would be possible to use the well-developed kinetic theory (see, for example, [5.7–5.9].

A promising approach of this sort was considered in [5.10]. Let a traveling quasimonochromatic light wave be absorbed in a resonance transition. Owing to the Doppler effect, only those particles will interact with the radiation which have their velocity projection v_x on the direction of the wave vector $\mathbf{k}$ close to the quantity Ω/k, where Ω is the wave frequency detuning from the central frequency of the absorption line having a near-Gaussian profile. In this case, it is essential that $\Omega \neq 0$. The excited particles will then acquire a directed velocity of $v_x = \Omega/k$, and in subsequent ionization (e. g., on account of an associative process the type of the Mölner-Hornbeck process or some other, see elsewhere), the resultant ions will retain approximately the same velocity, while the electrons, whose absolute velocity is obviously several orders of magnitude higher than Ω/k because of their small mass, will practically move isotropically while leaving for the walls. And it is the ion current that will determine the optogalvanic effect, the retarding bulk collisions being in a first approximation negligible.

The effect can most easily be described quantitatively for the case of plane geometry where the absorbing cell is a parallel-plate capacitor and the radiation propagates at right angles to the plates. Disregarding the effect of the space charge, one would write the following relation for the electric current flowing through the capacitor:

$$i = e \int dr \int_0^\infty \left[q(v_x, \mathbf{r}) - q(-v_x, \mathbf{r}) \right] dv_x, \qquad (5.1)$$

where $q(v_x, \mathbf{r})$ is the number of ionization events per unit time per unit volume occurring in a unit interval of the velocity projections v_x. According to [5.10], one can demonstrate that at not very high exciting laser radiation intensities in a rarefied gas medium Eq. (5.1) yields

$$i = eQ\xi(2/\pi) \arctan (\omega/\Gamma_0) \exp [-(\omega/k\mathbf{v})^2], \qquad (5.2)$$

where Q is the absorption rate of exciting radiation quanta in the cell, ξ is the ionization probability of the excited atoms, and $\mathbf{v}$ is their average thermal velocity. If the gap between the capacitor plates is substantially greater than the Debye radius, i. e., if the plasma conditions are satisfied, the following relation holds instead of Eq. (5.2):

$$i \cong eq_0 \sqrt{\left(T_{\mathrm{g}} / T_{\mathrm{e}}\right)} (\Omega / k\mathbf{v}) \exp\left[-(\Omega / k\mathbf{v})^2\right], \qquad (5.3)$$

where q_0 is the ionization rate. The functions $i(\Omega)$ calculated by formulas (5.2) and (5.3) are presented in Fig. 5.1. One can see that under free ion flight conditions the sign reversal of the current induced in the rarefied gas takes place within a very narrow transition region, $\Delta\Omega \sim \Gamma_0 \ll k\mathbf{v}$, i. e., the corresponding optogalvanic signal is, in principle, of sub-Doppler character. In the second case, the Coulomb interaction makes the frequency dpendence $i(\Omega)$ smoother, its sole characteristic scale being $\Omega/k\mathbf{v} \sim 1$.

The simple estimate by formula (5.2) of the current induced in the excitation region of the gas as a result of absorption of some 10^2 W of the exciting light, this corresponding to a thermal velocity of around 10^{17} s^{-1}, yields

$$i \sim 10^{-2}\xi \text{ A.} \qquad (5.4)$$

Present-day instruments are quite capable of measuring currents as small as 10^{-15} A, which means that the optogalvanic effect can be observed at ξ values up to some 10^{-13}. And the probability ξ is usually several orders of magnitude higher. For example, in the case of the associative ionization of sodium,

$$\text{Na}(3p) + \text{Na}(3p) \rightarrow \text{Na}_2^+ + e^-, \qquad (5.5)$$

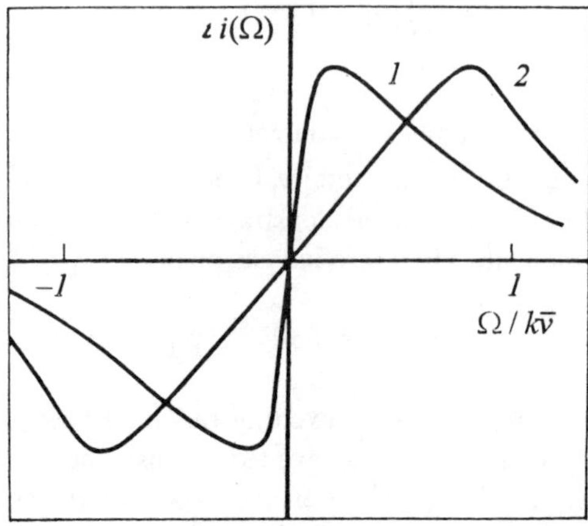

Figure 5.1. Functions $i(\Omega)$ calculated by formulas (5.2) (curve *1*) and (5.3) (*2*). *1* – free ion flight conditions, $\Gamma_0 / k\bar{v} = 10^{-2}$; *2* – plasma conditions [5.10].

substituting the values of the elementary process cross sections (known from the literature) and other characteristic quantities (resonance level lifetime, volume of the light-absorbing gas region, thermal velocities of particles, etc.) gives $i \sim 10^{-9}$ A.

In the experiment [5.10] involving the absorption of a dye laser radiation by Na vapor, special measures were taken to suppress the possible photoelectric effect caused by scattered light that could, in principle, distort the optogalvanic signal being detected. This experiment showed close agreement with theoretical estimates, and the optogalvanic effect was observed reliably when the laser frequency was varied symmetrically with respect to the central frequency of one of the D-lines of Na. The authors of [5.10] believe that the optogalvanic effect in a rarefied gas can form the basis for a new precision technique for high-resolution spectroscopy. Further investigations along these lines, both theoretical and applied, are undoubtedly desirable.

In light of this, the situation can be considered to be extreme. Of a more general character are the optogalvanic processes due to the light-induced e. m. f. in gases [5.11] and the light-induced current in weakly ionized plasmas [5.12].

If the atoms A contained in the buffer gas formed by the particles B can absorb in a resonance fashion the power of a traveling monochromatic wave, so that the Doppler broadening of the absorption line is predominant, the excited atoms A^*, A^{**}, and A^{***} (the double and triple asterisks denoting Rydberg states and states lying above the ionization limit, respectively) acquire a directional velocity. One can indicate several different pathways for subsequent ionization, the chief among them being as follows:

$$A^* + B \rightarrow (AB)^+ + e^-,$$

$$A^* + B \rightarrow A + B^+ + e^-,$$

$$A^{**} + B \rightarrow A^+ + B + e^-, \qquad (5.6)$$

$$A^{**} + B \rightarrow A^+ + B^-,$$

$$A^{***} \rightarrow A^+ + e^-$$

The first reaction in (5.6) is the above-mentioned associative ionization process which proceeds most effectively when the ionization potential of the excited atom A^* is commensurable with the dissociation energy of the resultant molecular ion $(AB)^+$ (the cross section of the process is approximately 10^{-15} cm^2). The second reaction corresponds to the case where the energy of the excited atom A^* exceeds the ionization potential of the atom B (a directional motion impulse is imparted to the ion B^+). The cross section values of the third and the fourth reaction are, as a rule, very great (10^{-11}–10^{-12} cm^2 [5.13]), so that the velocities of the colliding particles remain almost unchanged and the ions A^+ acquire the directional velocity of the Rydberg atoms A^{**}, while the ions B^- retain their isotropic velocity distribution. And finally, in the course of the fifth reaction in (5.6), whose products are ions and electrons, the charged particles differing in mass get separated in space because of the great difference between their retardation velocities.

The development of a light-induced e. m. f. in a gas or plasma is intimately connected with the selective diffusion or light-induced particle drift phenomenon that has been extensively studied in recent years [5.14, 5.15]. This connection is especially pronounced where the absorbents are ions. The pertinent experiments can be performed using both continuous-wave and pulsed lasers, the optimization of the optogalvanic experiments as to the pumping power being easier in the latter case.

We can easily write the following expressions for the laser-induced e. m. f. and the power P consumed from the current source [5.12]:

$$E = v_{\mathrm{d}}^{(i)} L / \mu_{\mathrm{e}}, \quad P = 2v_{\mathrm{d}}^{(i)2} eS_{\mathrm{L}} \Delta\omega / \left(\mu_{\mathrm{e}}\lambda^2 A_k\right), \tag{5.7}$$

where $v_{\mathrm{d}}^{(i)}$ is the ion drift velocity, L is the length of the absorbing cell, S_{L} is the cross-sectional area of the laser beam, λ is the wavelength of the laser light, A_k is the radiative decay rate of the kth state of the ion, and $\Delta\omega$ is the half-width of the absorption profile. Putting for estimation purposes $S_{\mathrm{L}} \cong 1$ cm^2, $\lambda = 0.3$ μm, $\mu_{\mathrm{e}} \cong 5$ m^2/V·s, and $\Delta\omega/A_k \cong 10^2$ and assuming that the ion drift velocity $v_{\mathrm{d}}^{(i)}$ ranges between 1 and 10% of the average thermal velocity $\bar{v}_{\mathrm{i}}$, we find that $P \cong 10^{-2}$–10^{-1} mW. A signal of so high a power can easily be detected against the discharge noise background.

Theoretically, the problem reduces to the correct calculation of the drift velocity $v_{\mathrm{d}}^{(i)}$. Generally speaking, the problem is not easy and requires, as in the case of the light-induced drift theory [5.16], consideration for a number of fine kinetic effects. But by using the popular strong-collision model [5.9], one can obtain a comparatively simple analytical expression for the velocity $v_{\mathrm{d}}^{(i)}$ (or directly for the ratio $v_{\mathrm{d}}^{(i)} / \bar{v}_{\mathrm{i}}$). According to [5.12], it has the form

$$\frac{v_{\mathrm{d}}^{(i)}}{\bar{v}_{\mathrm{i}}} = \frac{v_{\mathrm{n}} - v_k}{v_{\mathrm{n}} + v_k} \frac{\sqrt{\pi}\, y \operatorname{Re}\left[zf(z)\right]}{1 + 1/\kappa' + (\tau_2 / \tau_1)\sqrt{\pi}\, y \operatorname{Re}\left[f(z)\right]}, \tag{5.8}$$

where

$$f(z) = \exp\left(-z^2\right)\left(1 + \frac{2i}{\sqrt{\pi}} \int_0^z e^{t^2} dt\right), \quad z = x + iy;$$

$$x = \Omega / \left(k\bar{v}_{\mathrm{i}}\right), \quad y = \Gamma\left(1 + \kappa'\right)^{1/2} / k\bar{v}_{\mathrm{i}};$$

$$\tau_1 = \tau_{1_k} + \tau_{1_{\mathrm{n}}}, \quad \tau_2 = \tau_{2_k} + \tau_{2_{\mathrm{n}}};$$

$$\tau_{1_m} = \frac{1}{A_k + v_k}, \quad \tau_{2_m} = \frac{1}{A_k} - \tau_{1_k};$$

$$\tau_{1_{\mathrm{n}}} = \frac{v_m}{v_{\mathrm{n}}} \tau_{1_k}, \quad \tau_{2_{\mathrm{n}}} = \frac{1}{A_k} - \tau_{1_{\mathrm{n}}}.$$

In these formulas, Γ is the absorption profile half-width of an isolated ion, v_k and v_{n} are the frequencies of collisions between ions and neutral atoms, and κ' is the saturation parameter. The drift velocity vector $\mathbf{v}_{d=}^{(i)}$ is directed along the wave vector $\mathbf{k}$.

The analysis of expression (5.8) shows that if the optical-collisional exchange is important, the quantity τ_2 becomes negative, and as the ratio $|\tau_2|/\tau_1$ increases, the denominator in formula (5.8) diminishes. Satisfying the inequality $\nu_n \ll \Gamma_k$ by gradually reducing the buffer gas pressure and assuming that $\kappa' \gg 1$ and $\nu_k = 2\nu_n$, we find that the maximum ratio $v_d^{(i)} / \bar{v}_i$ corresponds to $x \cong y \cong 1$ and is

$$\left[v_d^{(i)} / \bar{v}_i \right]_{\max} \cong 0.06, \tag{5.9}$$

which agrees with the above estimate of the power P. Note the recent result of the experiment [5.17] on the light-induced drift of Na atoms in neon as a buffer gas: by using a silane coating on the absorption cell walls, the authors of this work have sharply reduced the role of the surface adsorption of the particles and thus managed to raise the light-induced drift velocity up to 1 m/s.

By reducing the buffer gas pressure further (the electron mobility μ_e in that case grows higher), one can make the rate of flux saturate at ν_k, $\nu_n \sim A_k$. To estimate the power P in this case, one can take $A_k \sim 10^8$ s^{-1} and $\Delta\omega \sim 10^{10}$ s^{-1}. This enables one to find out whether it is possible to conduct an optogalvanic experiment in the given weakly ionized gas conditions, and take precision absorption line profile measurements in particular [5.18].

The authors of [5.19] studied the generation of light-induced currents (LIC) in plasma conditions. The experiment was conducted in a glow discharge in hydrogen at a pressure of 0.3 Torr under irradiation at the wavelength of H_2. The magnitude of the light-induced current in these conditions (discharge current density of the order of 10 A/cm^2) was smaller by a factor of approximately 60 than that of the normal optogalvanic effect. However, these currents can be experimentally separated easily enough by varying the laser radiation frequency within the limits of the atomic line profile and illuminating the plasma alternately in opposite directions. In contrast to the optogalvanic signal, the sign of the LIC signal depends on the irradiation direction. The frequency dependence of the optogalvanic signal reproduces (see Sect 6.6) the line profile shape. And the LIC signal corresponds to the derivative of the profile, which is typical of the kinetic mechanisms of light-induced phenomena. This effect is in principle of individual interest, specifically in investigations into the nature of generation of magnetic fields in stellar atmospheres [5.20].

5.3. DOPPLER-FREE SATURATION AND TWO-PHOTON ABSORPTION OPTOGALVANIC SPECTROSCOPY

The high-resolution optogalvanic spectroscopy techniques considered in this section are direct evolution products of their purely optical analogs treated in detail in a number of reviews and monographs (see, for example, [5.2, 5.21–5.23].

Optogalvanic detection of the Lamb dip was carried out for the first time in 1978 by T. Johnson [5.24] who studied an absorption line profile in a He-Ne discharge. But the authors of [5.25] suggested a much more convenient saturation optogalvanic spectroscopy scheme on the basis of its fluorescence version proposed earlier in [5.26]. This scheme is known as IMOGS, from intermodulated optogalvanic spectroscopy.

The IMOGS technique, like its counterpart proposed in [5.26], is based on the effect of saturated absorption in the field of a standing or traveling light wave interacting with a gas (or plasma) medium (Fig. 5.2). The beam of a tunable dye laser is split into two components of approximately the same intensity, which are made to pass through the discharge in directions making an angle close to 180° with each other, so that these component beams overlap in some small volume of the discharge plasma. It is pertinent to note that the component beams used in the intermodulated fluorescence signal detection scheme [5.26] differed sharply in intensity (the so-called "exciting" and "probe" beams). By varying the angle between the beams slightly, one can change both the size of the beam overlapping region and the discharge region subject to the simultaneous action of both component beams.

The shutter serves to amplitude-modulate the beams with different frequencies, w_{1k} and w_{2k}, i. e.,

$$I_1 = I_{01}[1 + \cos(w_{1k}t)] \text{ and } I_2 = I_{02}[1 + \cos(w_{2k}t)], \tag{5.10}$$

the optogalvanic signal intensity being

$$\Delta i = Cn_s(I_1 + I_2), \tag{5.11}$$

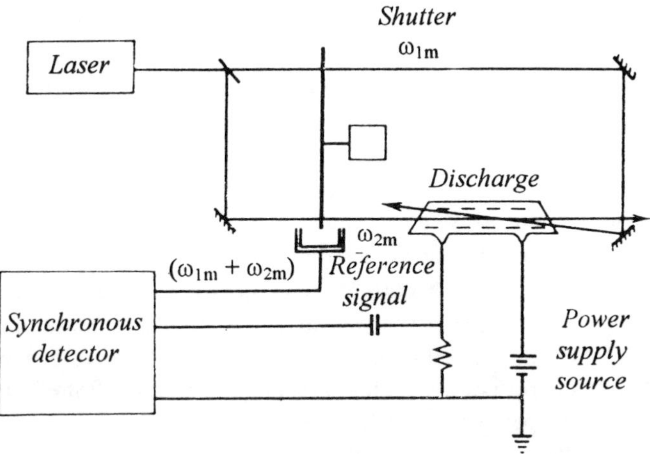

Figure 5.2. Schematic diagram of the experiment on detecting a Doppler-free intermodulated optogalvanic spectroscopy (IMOGS) signal.

where n_s is the saturated population in some section of the absorption line profile and C is the coefficient of proportionality. Evidently if the laser is not tuned exactly to the center of the Doppler-broadened line, each beam will only interact with a certain group (ensemble) of atoms whose velocity projections on the direction of the wave vector correspond to the resonance condition. In that case, a pair of independent optogalvanic signals modulated in amplitude at frequencies of ω_{1k} and ω_{2k} develops. If the laser frequency is tuned to the center of the Doppler profile (within the limits of the homogeneous half-width), both the above groups of atoms will merge into one wherein the atomic velocity projections on the direction defined by the beams are close to zero. In that case,

$$n_s = n_0(1 - \kappa_0') = n_0[1 + a(I_1 + I_2)] \tag{5.12}$$

(the explicit forms of the factors n_0 and a not important here), and according to Eq. (5.11),

$$\Delta i = Cn_0[(I_1 + I_2) - a(I_1 + I_2)^2]. \tag{5.13}$$

Thus, the optogalvanic signal Δi contains linear terms amplitude-modulated at the frequencies ω_{1k} and ω_{2k}, as well as quadratic terms modulated at the frequencies $2\omega_{1k}$ and $2\omega_{2k}$ and also at the sum and difference frequencies $(\omega_{1k} + \omega_{2k})$ and $(\omega_{1k} - \omega_{2k})$. The synchronous detector is usually tuned to the sum frequency $(\omega_{1k} + \omega_{2k})$ and the linear background is cut off, so that only the saturation signal is detected.

Figure 5.3 shows as an example a Doppler-free optogalvanic signal obtained with a setup using a helium glow discharge. The authors of [5.25] attained a high signal/noise ratio, the time constant being a mere 1 s. The modulation frequencies were selected to be $\omega_{1k} = 600$ Hz and $\omega_{2k} = 840$ Hz.

The background current in the frequency band $\Delta\omega \sim 1$ Hz at a sum frequency of $(\omega_{1k} + \omega_{2k}) = 1440$ Hz amounted to around 10^{-11} A, which corresponded to the characteristic values for the shot noise under the experimental conditions selected. This is at least two orders of magnitude lower than the corresponding values attainable within the framework of the traditional laser spectroscopy techniques [5.22].

More recently, other versions of the IMOGS technique were suggested,

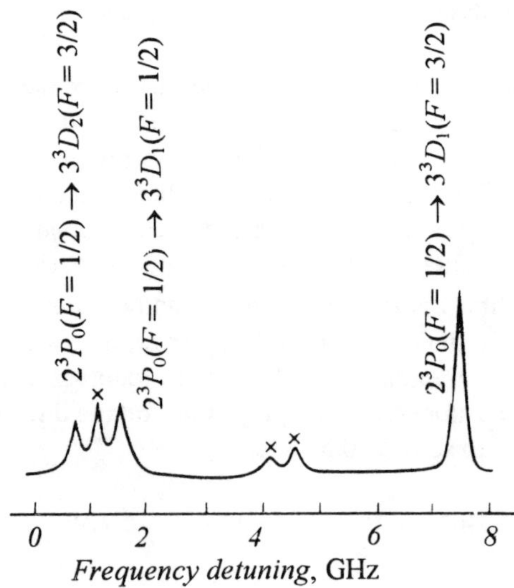

Figure 5.3. Intermodulated optogalvanic spectrum for a portion of the transition $2^3P \rightarrow 3^3D$ at a frequency of 587.5 nm in ^{3}He. The asterisks indicate the artifacts typical of all Doppler-free absorption saturation measurements.

which possessed a very high sensitivity, too, each being characterized by its own set of requirements for the optical elements, detectors, and charge parameters.

The two-photon optogalvanic spectroscopy idea, first implemented in [5.27], goes back to the work by Vasilenko *et al.* [5.28]. In the rest frame of reference of the absorbing atom (or molecule), account should be taken of the Doppler frequency shift of the electromagnetic wave with the wave vector $\mathbf{k}$:

$$\omega' = \omega - \mathbf{k}\mathbf{v}, \qquad (5.14)$$

where $\mathbf{v}$ is the atomic velocity in the laboratory frame of reference. In that case, if the two-photon resonance condition is satisfied for the absorbing transition $E_1 \rightarrow E_2$, i. e., if

$$(E_2 - E_1)/\hbar = \omega_1' + \omega_2' = \omega_1 + \omega_2 - \mathbf{v}(\mathbf{k}_1 + \mathbf{k}_2), \qquad (5.15)$$

it is easy to understand that in the case of counter-propagating waves of the same frequency ($\omega_1 = \omega_2$) the vector sum $\mathbf{k}_1 + \mathbf{k}_2 = 0$ and the Doppler shift does not show itself in any way. It can be seen that if a moving atom simultaneously interacts with more than two plane waves with the wave vectors $\mathbf{k}_i$, the Doppler shift for the transition as a whole will be nil, too, if the condition $\sum_i \mathbf{k}_i = 0$ is satisfied.

In practice, the optogalvanic detection of a two-photon resonance can most easily be implemented by passing a focused laser beam through a gas-discharge tube and then reflecting it with a mirror in the opposite direction. The optical system may also include a Faraday cell rotating the polarization plane through 45° and thus providing for optical decoupling [5.23]. The electrical circuit for detecting and recording the optogalvanic signal is a standard type.

In accordance with condition (5.15), when $\omega_1 = \omega_2$, it is necessary to attain resonance, by adjusting the laser frequency, with half the energy difference for the transition selected. The two-photon absorption process taking place case is described in the second-order perturbation theory approximation, the transition occurring between levels of the same parity.

When an atom or a molecule absorbs a photon from either of the two counter-propagating beams, the first-order Doppler effect will apparently be fully neutralized. Two-photon absorption from a single beam is possible in principle, but the signal corresponding to such a process would be weaker by several orders of magnitude than its Doppler-free counterpart, because it would only be contributed to by atoms with a strict limitation on their velocity projection. In the case of two beams propagating in the opposite directions (the slight deviations due to beam angle and overlapping region localization adjustments are quite admissible), all atoms will participate in the two-photon absorption process, no matter what their velocity.

The advantages of the optogalvanic detection technique in comparison with the traditional fluorescence methods are especially pronounced where transitions from excited levels that are not metastable are observed. Prior to the advent of the two-photon optogalvanic spectroscopy (TOGS) technique, there was actually not a single case where it would prove possible to reliably detect a two-photon absorption signal in a transition whose lower level was neither the ground, nor a metastable one.

The authors of [5.27] described the application of the TOGS technique to the solution of the intricate problem of the spectroscopy of the ^{20}Ne isotope excited by a dc-induced discharge in the tube of a commercial He-Ne laser (0.23 Torr ^{20}Ne + 1.62 Torr ^{3}He). Subject to study were four neon levels for the configuration $3s$. The discharge tube was placed directly inside the laser cavity, and use was made of frequency modulation instead of amplitude modulation. The synchronous detector used served to monitor the derivative of the Lorentzian-shape signal.

A typical signal recorded on the transition $3s'[1/2]$, $J = 1 \rightarrow 5d\,'[3/2]$, $J = 1$ with a time constant of the order of 1 s in ^{20}Ne is shown in Fig. 5.4. The level $3s'[1/2]$, $J = 1$ is not metastable, and its lifetime is around 1.5 ns. All in all, the authors managed to record 13 two-photon transitions, including those from all the four $3s$ levels (two of them are not metastable). In another work [5.29], the TOGS technique was used in a scheme with a narrow-band pulsed laser to detect the two-photon transitions $2^3s \rightarrow 5^3s$ and $2^3s \rightarrow 5^2D$ in ^{4}He. It was also found that by using the TOGS method with a two-photon transition with a high value of the oscillator force one can achieve very high signal/noise ratios.

5.4 NARROW OPTOGALVANIC RESONANCES WITH INTERMODULATED COUNTER-PROPAGATING BEAMS OF POLARIZED RADIATION

The POLINEX (polarized intermodulated excitation) technique developed comparatively recently by the authors of [5.30] is classed with the methods that allow for the optogalvanic detection of narrow nonlinear (Doppler-

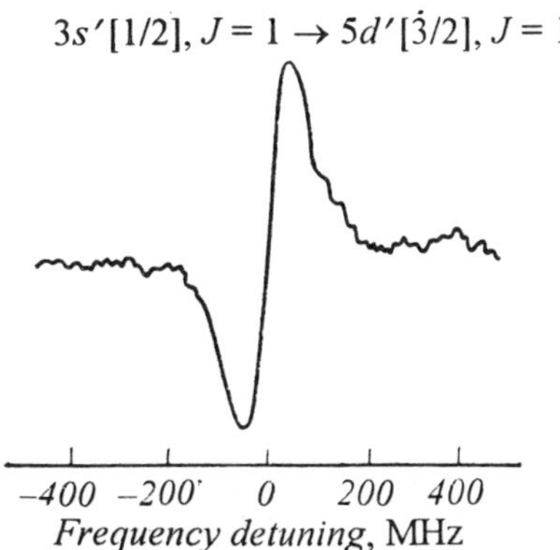

Figure 5.4. Nonlinear two-photon optogalvanic signal (TOGS) recorded on the transition $3s'[1/2]$, $J = 1 \rightarrow 5d'[3/2]$, $J = 1$ in ^{20}Ne.

free) resonances. This method is universal and, in addition to the optogalvanic technique, it can easily be combined with emission (fluorescence) spectroscopy and optoacoustic detection. As distinct from the above-mentioned intermodulated optogalvanic spectroscopy (IMOGS) technique based on the intensity-modulation of two counter-propagating laser beams, subject to modulation here are the polarization characteristics of either one of the two or both of the beams from a tunable laser that also propagate counter to each other and intersect within some limited region inside a gas discharge.

A typical experimental scheme is presented in Fig. 5.5. The polarization modulator may use a Pockels-effect cell or some other device with a birefringent ADP crystal, the modulation frequency (usually of the order of 1 kHz) being set by the same synchronous amplifier as forms part of the optogalvanic signal detection unit. The light beams interacting in the discharge plasma may have either linear or circular polarization.

The essence of the POLINEX technique may be revealed using two circularly polarized quasimonochromatic laser beams propagating counter to each other and interacting, as in the case of the IMOGS technique, with the atoms whose velocity projection on the wave vector direction is close to zero. If this direction is also the quantization axis and the influence of the external fields (a transverse electric or magnetic field) can be disregarded, either of the two beams will stimulate electric dipole transitions from the lower sublevel $|J, m\rangle$ to the upper one, $|J', m'\rangle$, where $m' = m + q$ and $q =$

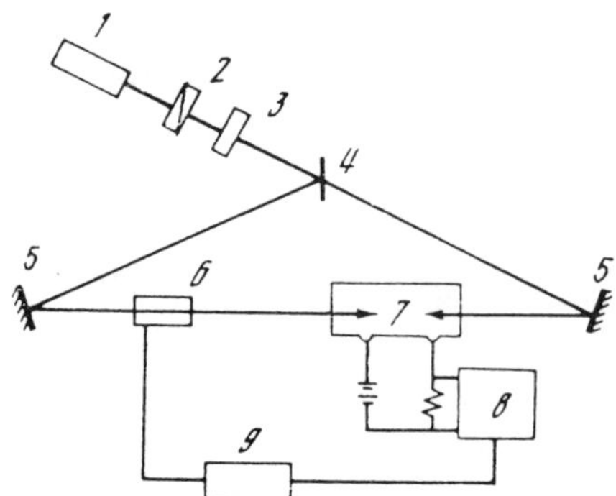

Figure 5.5. Schematic diagram of the experiment on detecting counter-propagating modulated beams of polarized radiation (POLINEX). *1* – laser; *2* – polarizer; *3* – quarter-wave plate; *4* – beam splitter; *5* – mirrors; *6* – polarization modulator; *7* – gas discharge; *8* – synchronous amplifier and detector; *9* – high-voltage amplifier.

± 1, depending on the circular polarization direction. According to the Wigner-Eckart theorem [5.31], the absorption cross sections $\sigma_{mm'}(q)$ in such transitions are proportional to the squares of the Clebsch-Gordan coefficients $\langle Jmlq|J'm'\rangle$. In particular, if $J' = J \pm 1$, those atoms will absorb the most, which have their angular momentum directed opposite to the circular polarization rotation vector. In other words, the maximum absorption occurs in the case of interaction of counter-propagating waves of the same circular polarization.

In that case, the population of the lower sublevel $|J, m\rangle$ of an atom subject to the action of two light waves with the intensities I_1 and I_2 and circular polarizations q_1 and q_2 may be expressed as

$$n_m = n - \sum_{m'} \frac{(n - n')}{\hbar\omega\,\Gamma}\left[I_1\sigma_{mm'}(q_1) + I_2\sigma_{mm'}(q_2)\right], \qquad (5.16)$$

where $n = N/(2J + 1)$ and $n' = N'/(2J' + 1)$ are the saturated equilibrium populations of the lower and upper states, respectively, their lifetimes being Γ and Γ', and N and N' are the total populations of the lower and the upper level. The expression for the population n' of the upper sublevel is written in the same way. The reabsorption and multiple-quantum pumping processes are disregarded here.

The structure of the expression for the total absorption coefficient k_0 in the given transition under saturation conditions is the same as in formula (5.13) pertaining to the IMOGS technique:

$$k_0 = \sum \frac{n - n'}{\hbar\omega} \left\{ I_1 \sigma_{mm'}(q_1) + I_2 \sigma_{mm'}(q_2) - \frac{1}{\hbar\omega}\left(\frac{1}{\Gamma} + \frac{1}{\Gamma'}\right) \times \right.$$
$$\left. \left[\left(I_1\sigma_{mm'}(q_1)\right)^2 + \left(I_2\sigma_{mm'}(q_2)\right)^2 + 2I_1I_2\sigma_{mm'}(q_1)\sigma_{mm'}(q_2)\right]\right\}. \tag{5.17}$$

As in formula (5.13), the first two linear terms characterize the saturated absorption in each of the beams, the two first quadratic terms (in square brackets) allow for the contribution due to the self-saturation of the beams, and finally the last term, which is due to intermodulation at low frequencies, describes the cross-saturation effect in the given transition. Let us emphasize that the optogalvanic signal contains no contribution from any of the above terms except the last one.

The optogalvanic signal is formed as a response to the difference Δk_0 between the absorption coefficient values in the case where the counter-propagating beams first have opposite and then the same circular polarization parameters. In that case,

$$\Delta k_0 = \frac{2I_1I_2}{(\hbar\omega)^2}\left(\frac{1}{\Gamma} + \frac{1}{\Gamma'}\right)(n - n')\sum_{mm'}\sigma_{mm'}(q_1)\sigma_{mm'}(q_2). \tag{5.18}$$

The sums of the products of the cross sections $\sigma_{mm'}$ entering into expression (5.18) can easily be tabulated for beams of both circular and linear polarization, as was done by Dabkiewicz and Hansch [5.32].

A distinctive feature of the POLINEX technique is that it takes advantage of the collisional depolarization process which is a parasitic factor in most of the other well-known high-spectral-resolution techniques. The atoms that change as a result of collision their velocity projection on the quantization axis (the latter usually coincides with the propagation axis of the beams) no longer participate in the formation of the optogalvanic signal and do not contribute anything undesirable to the pedestal above which the useful signal towers. In addition, the technique proves informative as to the angular momenta of the levels involved in the optical transition and the relaxation parameters of the light-induced atomic orientation and alignment [5.33, 5.34].

The authors of [5.30, 5.32, 5.35] experimentally studied the capabilities of the POLINEX technique using glow, high-frequency, and hollow-cathode discharges. The optogalvanic detection of narrow nonlinear resonances on the transitions HeI $2^3P \rightarrow 3^3D$ (λ = 587.5 nm), NeI $1s_5 \rightarrow 2p_2$ (λ = 588.2 nm), NeI $1s_5 \rightarrow 2p_4$ (λ = 594.5 nm), NeI $1s_2 \rightarrow 2P_1$ (λ = 585.2 nm), and CuI $3d^{10}4p^2P_{1/2} \rightarrow 3d^94s^{2\,2}D_{3/2}$ (λ = 578. 2 nm) fully attested to the merits of the technique. The use of this technique with electrodeless high-frequency discharges [5.35] ensures very high signal/noise ratios in conditions of virtually completely absent pedestal. The advantages of polarization modulation over its amplitude counterpart are evident from Fig. 5.6.

Similar results were later obtained by the authors of [5.36] who applied the POLINEX technique to the as yet little understood problem of hyperfine structure formation in atoms with an unfilled d-shell (e. g., molybdenum). The above authors described a sufficiently universal setup that allowed the implementation of various nonlinear Doppler-free spectroscopic techniques using both fluorescence and optogalvanic signal detection.

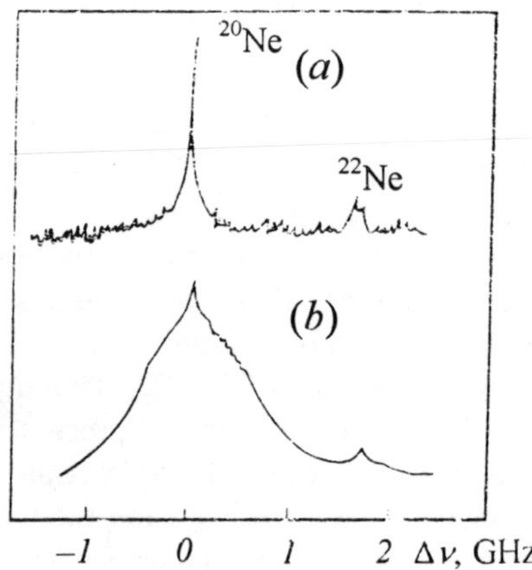

Figure 5.6. Doppler-free optogalvanic spectra of Ne (transition $1s_5 \rightarrow 2p_2$, λ = 588.2 nm) detected with electrodeless high-frequency discharge excitation [5.38]. (*a*) POLINEX technique; (*b*) IMOGS technique.

5.5 LEVEL-CROSSING OPTOGALVANIC SPECTROSCOPY

It is a well established fact that the change in the emission characteristics of atoms and molecules resulting from the degeneration of their levels (or the crossing of the latter has long since been successfully used to solve many precision problems in magnetometry, gyroscopy, and spectroscopy, such as the determination of hyperfine structure constants, Lande splitting factors, Stark shifts, radiative lifetimes, and so on [5.37]. The majority of applications are associated with the frequently occurring case of level crossing in a zero magnetic field, the so-called Hanle effect discovered even before the advent of quantum mechanics [5.38].

The method of studying level-crossing effects on the basis of fluorescence spectra has recently been well developed. At first glance, the use of fluorescence spectra seems preferable because they directly reflect such processes as the alignment or orientation of atoms and molecules. At the same time, optogalvanic signals that primarily reflect level population changes are, to a first approximation, customarily considered to be isotropic, insensitive to the polarization characteristics of the ensemble of particles.

Nevertheless, a number of interesting possibilities have recently been revealed for the optogalvanic detection of level-crossing effects. First, as indicated in [5.33, 5.39, 5.40], a relationship exists between the ionization of excited atoms and their degree of alignment. The authors of [5.41] verified this relationship experimentally by placing neon gas-discharge lamps and thermionic diodes in an external magnetic field around 0.1 T in strength. Though the verification by the above authors of the theoretical notions developed in [5.33, 5.39, 5.40] was indirect (they detected the phase shift of the π- and σ-components at various external field strength values), there are strong grounds to believe that the predicted effects were, in fact, verified.

Secondly, the manifestation of degenerate states via their populations can also be successfully detected by the optogalvanic technique. In that case, the laser pumping of the selected spectral transition must provide for conditions close to saturation. The pertinent theoretical aspects of the problem were analyzed in detail in [5.42, 5.43]. We will refrain here from touching upon the fairly trivial case where the Zeeman sublevel splitting is comparable with the Doppler half-width and the disturbance of the resonance reduces the absorption of the pumping laser radiation.

Much more interesting is the situation where the external magnetic field is comparatively weak, so that the Zeeman splitting $\Delta\omega_p$ proves to be of the order of the inverse coherence relaxation time $2\Gamma_{ab}$ and then decreases progressively to lead eventually to the case of purely degenerate system (Fig. 5.7). Following the simple and graphic model suggested in [5.44], consider the transition between the levels a and b with the total angular momenta $J_a = 0$ and $J_b = 1$, the pumping laser radiation being linearly polarized normal to the direction of the magnetic field H splitting the upper level into three magnetic sublevels $m_b = 0, \pm 1$. The collisional relaxation among the sublevels is taken to be negligible.

Next compare the following three cases: (a) $\Delta\omega_p > \Gamma_{ab}$, the laser radiation interacts with two groups of atoms differing in velocity, and the population of the state J_b under saturation conditions is equal to $n/2 + n/2 = n$; (b) $\Gamma_{ab} > \Delta\omega_p > \Gamma_b$ (Γ_b is the homogeneous half-width of the upper level), the radiation interacts with a single group of n atoms, the populations of the magnetic sublevels $m_a = 0$ mad $m_b = +1$, and $m_b = -1$ are equalized so that the population of the state J_b is equal to $n/3 + n/3 =$

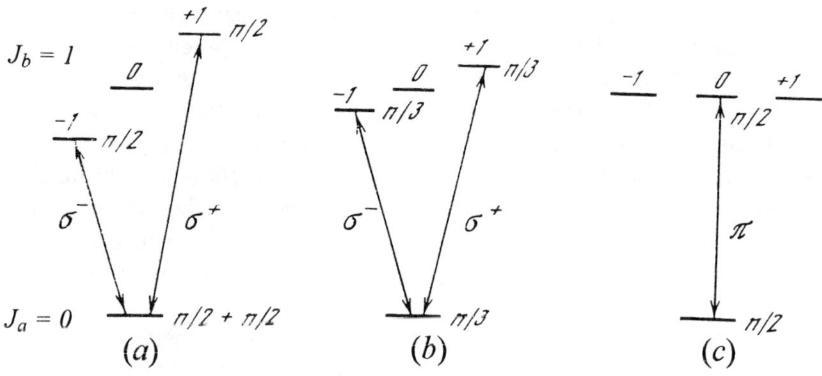

Figure 5.7. Various cases of the removal and recovery of the degeneration of the level $J_b = 1$. (a) $\Gamma_{ab} < \Delta\omega_p$; (b) $\Gamma_b < \Delta\omega_p < \Gamma_{ab}$; (c) $\Delta\omega_p = 0$. See text for explanations.

$2n/3$; (c) $\Delta\omega_p = 0$, there takes place the coherent superposition of states, so that the σ^+- and σ^-- transitions are recognized as a single π-transition and the population of the state J_b is obviously equal to $n/2$. Thus, when the magnetic field is varied, the resultant change in the population of the upper level should be described by the convolution of two Lorentzians with the half-widths Γ_{ab} and Γ_b. In typical cases, Γ_{ab} amounts to approximately 1% of the Doppler half-width, Γ_b being another order of magnitude narrower. In the model under consideration, the relative widths of these resonances are $S_1 = (n - 2n/3)/n = 0.33$ and $S_2 = (2n/3 - n/2)/(2n/3) = 0.25$.

To go over from populations changing in line with the magnetic filed variations to the characteristics of the actual optogalvanic signal, it is necessary to use a gas-discharge model. In general, it is important to take into consideration the possibility of the collisional mixing of the sublevels m_b, incomplete saturation on the line wings, radiation trapping, the effect of the optical pumping onto sublevels not directly in resonance, etc. All of these questions still need to be theoretically analyzed. Note that the methods to allow for radiation trapping in the case of linear Hanle effect were examined in [5.45].

As indicated in [5.42, 5.43], the main results of the theoretical analysis of the shape of the optogalvanic signal in conditions of magnetic field variations are confirmed by experiment. The authors of [5.44] described a series of optogalvanic spectroscopy investigations into the crossing of atomic levels in both zero and nonzero fields. The object of study was a low-pressure Ne or Ar discharge with an additional electrode inserted into the discharge plasma to atomize the element of interest (zirconium, yttrium). As a result, they demonstrated a number of new capabilities of the degenerate-state interference techniques. Among these are the determination of low values of hyperfine splitting constants (magnitude and sign) and studies into the characteristics of very weak transitions and Rydberg states formerly beyond the reach of the traditional optical detection methods. They also showed that optogalvanic techniques are not subject to stringent requirements on the stability of the laser frequency and ways to vary it.

Strumia [5.43] obtained interesting results concerning the mechanism responsible for the enhancement of the lasing power of continuous-wave argon- and krypton-ion lasers placed in an external axial magnetic field with a strength of some 0.1 T. The author suggested an unusual explanation of the enhancement of the lasing power in individual lines, which associated it not with the increase of the electron density of the plasma, but with the nonlinear Hanle effect. Qualitative agreement was noted between experimental and theoretical results for the power output in various spectral lines. It should be noted that current, more detailed studies require that the laser should operate under single-mode conditions.

5.6 STATE-SELECTIVE OPTOGALVANIC DOUBLE-RESONANCE SPECTROSCOPY

Although the problem of producing and detecting narrow nonlinear resonances is very important in high-resolution spectroscopy, another important concern in experimental practice is the capacity of the detection method used to identify or artificially simplify a complex absorption (or emission) spectrum. By "simplification" we mean the isolation (selection) of individual spectral transitions and elimination of the rest of the complex spectrum. A number of suitable techniques were described in [5.2]; note also the very important work by Wieman and Hänsch [5.46] on laser-induced polarization spectroscopy that gave impetus to rapid progress in this field.

But the direct extension of the principles underlying the purely optical selection of spectral transitions in atoms and molecules to the optogalvanic detection techniques usually proves difficult. An original way to solve such problems was suggested by Vidal [5.47], now known as optogalvanic double-resonance (OGDR) spectroscopy.

The Vidal scheme presented in Fig. 5.8 resembles the IMOGS scheme (cf. Fig. 5.2). The difference, however, is that use is made of two lasers operating at frequencies ω_1 and ω_2; the laser beams are then modulated in intensity with frequencies ω_{1m} and ω_{2m}, respectively. The changes in the discharge characteristics as a result of the action of the lasers are detected by a synchronous detector either at the difference frequency $|\omega_{1m} - \omega_{2m}|$ or at the sum frequency $(\omega_{1m} + \omega_{2m})$.

Typical features of the OGDR spectroscopy technique can easily be understood with reference to Fig. 5.9. If, for example, the laser frequency ω_1 is in resonance with the transition frequency ω_{AB}, the degree of plasma ionization associated with the populations n_A and n_B will be frequency-modulated at a frequency of ω_{1m}. A similar conclusion can be drawn about the frequency ω_2 of the second laser. When one of the levels proves to be common to both transitions (the level A in Fig. 5.9a, or the level B in Fig. 5.9b), it is precisely the effect of this level that will determine the shape and intensity of the optogalvanic signal detected at a frequency of

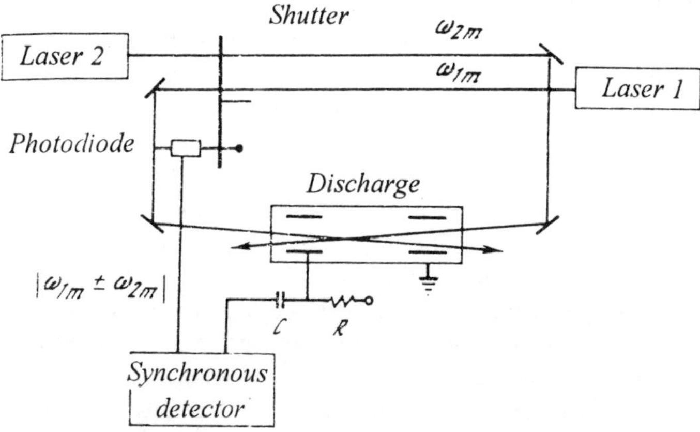

Figure 5.8. Schematic diagram of an experiment on selective optogalvanic double-resonance spectroscopy.

$|\omega_{1m} - \omega_{2m}|$ or $(\omega_{1m} + \omega_{2m})$. The necessary selection of transitions is thus ensured in the case of complex spectra. However, as distinct from the purely optical methods [5.2, 5.46], this technique provides for a substantial improvement of the signal/noise ratio and removes restrictions associated with the exclusive use of the electronic ground state (in complex molecular spectra).

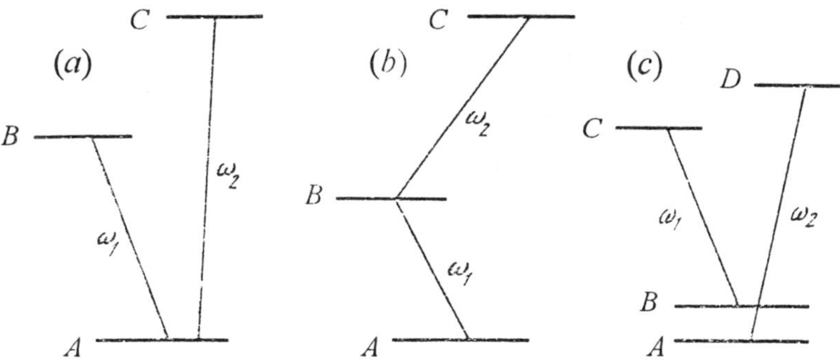

Figure 5.9. Examples of three- and four-level selective optogalvanic double-resonance spectroscopy schemes.

Another interesting possibility is illustrated by the schematic diagram shown in Fig. 5.9 c. The modulation of the populations of the levels B and C at a frequency of ω_{1m} may also manifest itself on the neighboring states, for example, A and D. In that case, by making the second laser frequency ω_2 resonate with the frequency of the transition $A \to D$, one can detect at the difference or sum frequency optogalvanic satellites carrying important information on the effective collision cross sections that can be extracted.

And finally, as regards principles, there are many possibilities of using optogalvanic double-resonance spectroscopy in conjunction with the narrow nonlinear resonance technique.

The work by Engelman and Keller [5.48] may be considered as the first experimental investigation into optogalvanic double-resonance spectroscopy wherein radiation from two modulated lasers was made to enter in the form of weakly convergent light beams a hollow-cathode discharge. The effect was observed in neon and also in sodium and uranium vapors, one of the lasers being modulated with a frequency of 28 Hz and the other, 2000 Hz or operated in a continuous-wave mode without any modulation at all. The amplification of the resultant signal with two resonances having a common intermediate level (Fig. 5.9c) being attained was so high that there was no need to carry out detection at either the difference or the sum modulation frequency.

A series of interesting OGDR spectroscopy experiments were conducted in [5.49–5.51]. The objects of study were neon, molecular hydrogen, and molecular nitrogen. In the latter case, a number of bands in the $b'^1\Sigma_u^+ \to a''^1\Sigma_u^+$ system were identified for the first time, thanks to the resonance selection of the transitions. As in [5.48], use was made of a hollow-cathode discharge, but double resonances were detected within a limited region of space in convergent counter-propagating laser beams.

The authors of [5.50, 5.51] developed a simple linearized model for calculating the most important characteristics of the optogalvanic signal detected in double-resonance experimental conditions.

5.7 MULTIPLE-QUANTUM AND STEPWISE EXCITATION

In the case of high radiation power, multiple-photon excitation may contribute to the optogalvanic effect. This may be most important for particles with a high ionization potential. Such a mechanism was observed, for example, in a flame [5.52]. The probability of the process depends, in particular, on the proximity of the actual and the intermediate virtual level. To illustrate, the detection of the two-photon effect in absorption at the $He(2^3S)$ levels requires much higher powers than in the case of absorption at the $Ne(3s)$ levels. In helium, the energy interval between the actual and the virtual level is 10^2 times greater than in neon, which lowers the probability of the two-photon process by a factor of around 10^4.

In strong light fields, resonance multiphoton ionization may take place. This process, detected by monitoring changes in the plasma conductivity, was observed in [5.53] when detecting oxygen atoms in hydrogen and also in [5.54] where PO and NO molecules were detected.

The use of two lasers to effect a stepwise excitation via an actual level requires much lower powers [5.55–5.57]. Concurrent irradiation on two resonance transitions with a common level may cause a substantial enhancement of the optogalvanic effect in comparison with cases where irradiation is carried out at either of these frequencies separately (double resonance).

Table 5.1 compares between the detection sensitivities for a number of elements in an air-hydrogen flame, attained with single- and

Table 5.1. Enhancement of the detection sensitivity in two-step excitation [5.55].

Element	E_i, cm^{-1}	E_0, cm^{-1}	λ_1, nm	E_1, cm^{-1}	p_1, µJ	S_1, nA/ppb
Ca	49 305	0	422.7	23 652	14	3×10^{-4}
Cu	62 317	11 203	510.6	30 784	150	10^{-5}
Fe	63 700	7 377	364.8	34 782	180	4×10^{-3}
In	46 670	2 213	451.1	24 373	120	0.06
Li	43 487	0	670.8	14 904	160	1.9
Mn	59 970	0	403.1	24 802	230	0.25
Na	41 449	0	589	16 973	26	1.5

Element	λ_2, nm	E_2, cm^{-1}	p_2, µJ	S_{1+2}, nA/ppb	S_{1+2}/S_1
Ca	518.9	42 919	75	0.08	270
Cu	453.1	52 849	110	0.016	1600
Fe	538.3	53 353	17	0.007	18
In	501.8	44 294	300	15	250
Li	610.4	31 283	40	195	100
Mn	602.2	41 404	33	1.7	7
Na	568.8	34 549	56	1000	690

two-step excitation schemes [5.55]. The detection sensitivity S is measured in units of current per particle surrounded by 10^9 solvent particles (nA/ppb). The lasers operate in a pulsed mode at wavelengths of λ_1 and λ_2 and with pulse energies of p_1 and p_2; E is the ionization energy, E_0 is the energy of the initial state for the excitation of the state with the energy E_1 by radiation of wavelength λ_1, and E_2 is the energy of the state excited from the state with the energy E_1 by radiation of wavelength λ_2. The enhancement of the detection sensitivity in the case of two-frequency excitation amounts to some $10–10^3$. It is important to note that the technique provides for the locality of the region of space wherein the optogalvanic signal is being generated on condition that the laser beams in the object of study are noncollinear. In that case, though, the question remains (as with other local optogalvanic spectroscopy modifications) as to how this locality affects the formation of the optogalvanic signal from the object as a whole. This question requires individual investigations for it to be answered. Emphasis was placed in [5.47, 5.51] on the frequency selectivity of double optogalvanic resonance, which is important in studying complex spectra. Also considered in the same works were various versions of the technique, including those cases where the common level was either the upper or the lower one, as well as the case where two optical transitions involved states close in energy and coupled by collisional exchange. Such a modification may prove useful in studying relaxation processes. The authors of [5.58] investigated the optogalvanic effect resonance in the case of concurrent irradiation of a plasma with a laser and microwave radiation (optical-microwave resonance).

REFERENCES

5.1. V. S. Letokhov, *Lazernaya fotoionizatsionnaya spektroskopiya* Laser photoionization spectroscopy) (Moscow: Nauka, 1987) (in Russian).

5.2. V. Demtryoder, *Lazernaya spektroskopiya: Osnovnye printsipy i tekhnika eksperimenta* (Laser spectroscopy: Basic principle and experimental techniques) (Moscow: Nauka, 1968) (in Russian).

5.3. J. Herrman, B. Wilhelmi, *Laser für Ultrakurze Lichtimpulse* (Berlin: Akademie-Verlag, 1984).

5.4. A. E. Bulyshev, N. G. Preobrazhensky, in: *Vsesoyuznyi simpozium po vychislitelnoi tomografii* (All-Union symposium on computational tomography) (Novosibirsk, 1983): 16–17 (in Russian).

5.5. A. Wexler, B. Fry, M. Neuman, *Appl. Optics* **24**: 3985–3992 (1985).

5.6. A. E. Bulyshev, N. V. Denisova, N. G. Preobrazhensky, in: *Materialy Vsesoyuz. seminara. Tallinn, 5–7 aprelya 1988* (Proceedings of the All-Union seminar) (Tallinn, April 5–7, 1988): 44–45 (in Russian).

5.7. J. Ferziger, H. Kaper, *Mathematical Theory of Transport Processes in Gases* (Amsterdam: North-Holland, 1972).

5.8. E. M. Lifshits, L. P. Potayevsky, *Fizicheskaya kinetika* (Physical kinetics) (Moscow: Nauka, 1979) (in Russian).

5.9. S. G. Rautian, G. I. Smirnov, A.M. Shalagin, *Nelineinye rezonansy v spektrakh atomov i molekul* (Nonlinear resonances in atomic and molecular spectra) (Novosibirsk: Nauka, 1979) (in Russian).

5.10. S. N. Atutov, I. M. Yermolayev, A. M. Shalagin, *Pis'ma v ZhETF* **40**, No. 9: 374–377 (1984).

5.11. A. I. Parkhomenko, V. E. Prpkop'yev, *Optika i Spektroskopiya* **53**, No. 6: 1000–1004 (1982).

5.12. F. Kh. Gel'mukhanov, A. M. Shalagin, *Kvantovaya Elektronika* **8**, No. 3: 590–594 (1981).

5.13. V. B. Leonas, in: *Itogi nauki i tekhniki* (Results of the development of science and technology) (Moscow: VINITI AN SSSR, 1980): 231 (in Russian).

5.14. F. Kh. Gel'mukhanov, A. M. Shalagin,, *Pis'ma v ZhETF* **29**: 773–776 (1979).

5.15. V. R. Mironenko, A. M. Shalagin, *Izv. AN SSSR. Ser. Fiz.* **45**: 995–1006 (1981).

5.16. B. Ya. Dubetsky, *ZhETF* **88**, No. 5: 1586–1599 (1985).

5.17. J. H. Xu, M. Allegrini, S. Gozzini, *Opt. Commun.* **63**: 43–48 (1987).

5.18. V. S. Kalinov, S. I. Ovseichuk, *Zhurn. Prikl. Spektroskopii* **47**, No. 2: 195–199 (1987).

5.19. S. N. Atutov, K. A. Nasyrov, S. P. Pod'jachev, A. M. Shalagin, *Phys. Rev. Lett.* **72**, No. 23: 3654–3657 (1994).

5.20. K. A. Nasyrov, A. M. Shalagin, *Astron. Astrophys.* **268**: 201–206 (1993).

5.21. V. S. Letokhov, V. P. Chebotayev, *Printsipy nelineinoi lazernoi spektroskopii* (Principles of nonlinear laser spectroscopy) (Moscow: Nauka, 1975) (in Russian).

5.22. V. S. Letokhov, in: K Shimoda, ed., *Topics in Applied Physics* (Berlin: Springer-Verlag, 1970), vol. 13: 95–171.

5.23. N. Bloembergen, M. D. Levenson,: in: K Shimoda, ed., *Topics in Applied Physics* (Berlin: Springer-Verlag, 1970), vol. 13: 315–369.

5.24. T. F. Johnson, *Laser Focus* (March 1978): 58–63.

5.25. J. E. Lawler, A. I. Ferguson, J. E. M. Goldsmith, *Phys. Rev. Lett.* **42**: 1046–1049 (1979).

5.26. M. S. Screm, A. L. Schawlow, *Opt. Commun.* **5**: 148–151 (1972).

5.27. J. E. M. Goldsmith, A. I. Ferguson, J. E. Lawler, A. L. Schawlow, *Optics Lett.* **4**: 230–232 (1979).

5.28. L. S. Vasilenko, V. P. Chebotayev, A. V. Shishayev, *Pis'ma v ZhETF* **12**, No. 3: 161–165 (1970).

5.29. J. E. M. Goldsmith, A. V. Smith, *Opt. Commun.* **32**: 403–405 (1980).

5.30. T. W. Hänsch, D. R. Lyons, A. L. Schawlow, *Opt. Commun.* **38**: 47–51 (1981).

5.31. V. B. Berestestky, E. M. Lifshits, L. P. Pitayevsky, *Kvantovaya Elektronika* (Quantum Electronics) (Moscow: Nauka, 1980) (in Russian).

5.32. Ph. Dabkiewicz, T. W. Hänsch, *Opt. Commun.* **38**: 351–356 (1981).

5.33. L. Julien, M. Pinard, *J. Phys. Atom. Molec. Phys. B* **15**: 2881–2889 (1982).

5.34. M. Pinard, L. Julien, *J. de Phys.* **44**, C7: 129–136 (1983).

5.35. D. R. Lyons, A. L. Schawlow, G. Y. Yan, *Opt. Commun.* **38**: 35–38 (1981).

5.36. Ch. Belfrage, P. Grafström, S. Kröll, *et al.*, *J. de Phys.* **44**, C7: 169–174 (1983).

5.37. M. P. Chaika, *Interferentsiya vyrozhdennykh atomnykh sostoyany* (Interference of degenerate atomic states) (Leningrad: Izd.-vo LGU, 1975) (in Russian).

5.38. L. N. Novikov, G. V. Skrotsky, G. I. Solomakho, *UFN* **113**: 597–625 (1947).

5.39. G. W. Series, *Comments Atom. Mol. Phys.* **10**: 199–201 (1981).

5.40. A. G. Petrashen', V. N. Rebane, T. K. Rebane, *Optika i Spektroskopiya* **57**: 963–964 (1984).

5.41. L. R. Pendrill, M. Petterson, U. Osterberg, *J. de Phys.* **44**, C7: 489–496 (1983).

5.42. J. N. Dodd, *J. Phys.* **16**: 2721–2731 (1983).

5.43. F. Strumia, *J. de Phys.* **44**, C7: 117–126 (1983).

5.44. P. Hannaford, D. S. Gough, G. W. Series, *J. de Phys.* **44**, C7: 107–115 (1983).

5.45. N. G. Preobrazhensky, A. E. Suvorov, *Effekt Khanle v poluprozrachnom gaze* (The Hanle effect in a semitransparent gas) (Novosibirsk: Preprint No. 2 In.-ta Teoret. i Prikl. Mekhaniki SO AN SSSR, 1978) (in Russian).

5.46. C. Wieman, T. W. Hänsch, *Phys. Rev. Lett.* **36**: 1170 (1976).

5.47. C. R. Vidal, *Optics. Lett.* **5**: 158–159 (1980).

5.48. R. Engelman, R. A. Keler, *Optics. Lett.* **5**: 465–466 (1980).

5.49. K. Miyazaki, H. Scheingraber, C. R. Vidal, *Phys. Rev. Lett.* **50**: 1046–1049 (1983).

5.50. K. Miyazaki, H. Scheingraber, C. R. Vidal, *J. de Phys.* **44**, C7: 411–418 (1983).

5.51. K. Miyazaki, H. Scheingraber, C. R. Vidal, *Phys. Rev.* **28**: 2229–2244 (1983).

5.52. G. C. Turk, J. C. Travis, J. R. De Voe, T. C. O'Haver, *Analyt. Chem.* **51**: 1890–1896 (1979).

5.53. J. E. M. Goldsmith, *Opt. Lett.* **7**: 437–439 (1982).

5.54. K. S. Smith, W. G. Mallard, *J. Chem. Phys.* **77**: 1779–1787 (1982).

5.55. N. B. Zorov, Yu. Ya. Kuzyakov, O. I. Matveyev, *Zhurn. Analit. Khimii* **37**: 520–533 (1982).

5.56. G. C. Turk, W.G. Mallard, P. K. Schenk, K. S. Smyth, *Analyt. Chem.* **51**: 2408 (1979).

5.57. P. Camus, M. Dieulin, El. Himdy, M. Aymar, *Phys. Scripta* **27**: 125–159 (1983).

5.58. C. Hameau, J. Wascat, D. Dangoisse, P. Glorieux, *Opt. Commun.* **49**: 423–428 (1984).

Chapter 6

Some Applications of the Optogalvanic Effect

6.1 ATOMIC SPECTROSCOPY

Optogalvanic spectra for many elements have been studied (see Table 6.1). The data listed in the table were borrowed from [6.1] and supplemented by us (see also [6.2]).

Apart from its purely applied aspects (diagnostics, analysis, etc.), the optogalvanic effect has found use in the study of atomic structures. The high sensitivity of the optogalvanic technique has determined the trend of these investigations, namely, the spectroscopy of excited states, small impurities, ions, low-probability transitions, and the like, where the common absorption and emission measurements prove difficult to take because of the low signal/noise ratio values. We will restrict ourselves here to a few examples.

Camus and co-workers [6.3] investigated the state spectrum of the barium atom in the energy range 5.2–7 eV. They identified 636 new levels with the principal quantum number up to $n = 45$ with $J = 0$–5 and determined their energies, the hyperfine structure being resolved for levels with $n \leq 25$. Rydberg-state spectra were studied in alkali metals. The energies and isotope shifts of levels were determined in $^{39, 40, 41}$K [6.4–6.7], $^{6, 7}$Li [6.8], ^{85}Rb [6.4, 6.9], and ^{133}Cs [6.10]. An essentially nonmonotonic relationship was found to exist between the quantum defect and the number

Table 6.1. Elements whose spectra were studied by optogalvanic spectroscopy techniques.

H (1, 5)							He (1, 3)		
Li (3, 5)	Be (4)	B (5)			O (5)		Ne (1–3)		
Na (3–6)	Mg (4, 5)	Al (5)					Ar (1–3)		
K (3–5)	Ca (3–5)	Sc (5)	Ti (5)	V (5)	Cr (5)	Mn (3–5)	Fe (5)	Co (3, 5)	Ni (5)
Cu (3–5)		Ga (3–5)					Kr (1, 2)		
Rb (4, 5)	Sr (1–3)	Y (3, 5)	Zr (3)		Mo (3)				
Ag (5)	Cd (5)	In (3)	Sn (5)				Xe (2)		
Cs (1, 2) (4, 5)	Ba (1, 3) (5)	La (3)							
Au (5)	Hg (1)	Tl (5)	Pb (5)	Bi (5)					
					U (3)				
Sm (3)			Tu (5)	Yb (3, 4)	Lu (5)				

Note. The figures in parentheses indicate the object wherein the optogalvanic effect was studied: 1 – dc-excited discharge; 2 – high-frequency discharge; 3 – hollow-cathode discharge; 4 – thermionic diode; 5 – flame; 6 – rarefied gas.

n in the nD_j series in cesium [6.10]. The isotope level shifts in [88, 86]Sr were studied in both neutral and ionized atoms [6.11].

Roesch [6.12] suggested using the high signal/noise ratio typical of optogalvanic spectroscopy even with very weak transitions to detect the parity violation effect on the $6S \rightarrow 7S$ transition in Cs.

The authors of [6.13, 6.14] investigated the optogalvanic spectrum of uranium in a hollow-cathode discharge. Despite the low concentration of [235]U (0.3%), its spectra proved to be reliably detectable. The optogalvanic effect can also be observed on the UII transitions. Figure 6.1 presents the

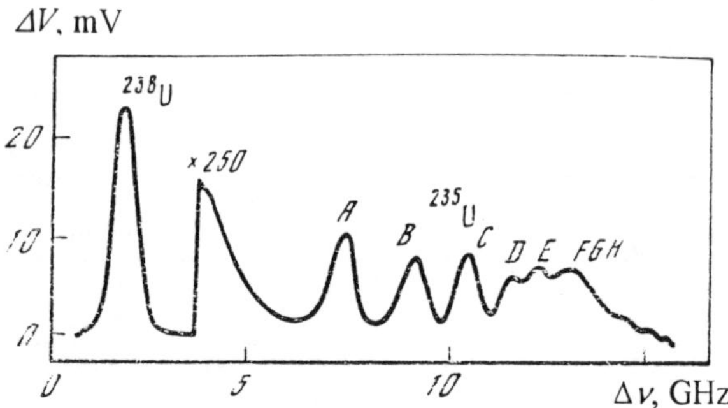

Figure 6.1. Optogalvanic spectrum of uranium [6.14] (the letters indicate the hyperfine-structure components of ^{235}U).

spectrum of uranium at approximately 591.5 nm, with resolved hyperfine-structure components and the ^{235}U–^{238}U isotope shift. Note also that a special hollow cathode design was proposed in [6.15] to study the optogalvanic spectra of difficult-to-evaporate elements with a resolution not limited by the Doppler broadening.

As the principal quantum number increases, level perturbation effects in the interaction between states converging to different ionization limits start to manifest themselves. Measurements taken for an extensive series of levels make it possible to compare between the data obtained and the existing theories. For example, Rinneberg and Neukammer [6.16] noted a discrepancy between the experimental data for Ba and the results of the multichannel quantum defect theory. The perturbation of states strongly affects the character of the hyperfine structure. Measurements are mainly taken with a resolution not limited by the Doppler broadening with a view to determining hyperfine structure and level mixing constants (the first investigations of the hyperfine structure of the $2^3P \rightarrow 3^3D$ transition in ^{3}He by an optogalvanic technique were conducted in [6.17]). As high-lying levels draw closer to one another, interaction occurs between multiplets having different principal quantum numbers but equal total momenta (n-mixing). The effect becomes noticeable at $n \geq 100$, and it was detected experimentally on an optogalvanic spectrum for $n \geq 90$ [6.18] (subject to detection were the levels with $n \leq 160$). The authors of [6.19, 6.20] studied the diamagnetic behavior of Rydberg states in external fields where the Lorentz force acting upon the outer-shell electrons became comparable with the Coulomb interaction force between the electrons and the atomic skeleton.

Some applications of the optogalvanic effect in polarization spectroscopy have already been considered earlier.

Optogalvanic techniques can help one to study autoionization states [6.21, 6.22, 6.23].

6.2 MOLECULAR SPECTROSCOPY

Although most publications deal with atomic spectra, a substantial number have also appeared on the studies of optogalvanic molecular spectra as well. The main trends in these investigations have recently been reviewed in [6.24, 6.25]. Table 6.2 summarizes and supplements the results presented in the above reviews.

Table 6.2. Studies of optogalvanic molecular spectra.

Molecule	Spectral range, nm	Reference	Note
Cs2	620–650	[6.31]	Diode. First observations of optogalvanic molecular spectra.
In$_2$ Yb$_2$ CsKr CsAr	390–660	[6.32–6.36]	Diode. Electronic transitions in excimers and exciplexes. Hybrid resonances due to atoms excited via intermediate dissociative molecular states.
LaO YO ScO	360–630	[6.37]	Flame. Identification of new transitions.
N$_2$	536–615	[6.38]	DC-excited discharge. 1^+ system and the Ledbetter bands ($c'^1\Pi \rightarrow a''^1\Sigma$).
N$_2$	598	[6.28]	High-frequency discharge. Doppler-free resolution.
N$_2$	585–605	[6.28, 6.39]	High-frequency discharge. Identification of new bands.
N$_2$	595–615	[6.27]	Hollow-cathode discharge. Double optical resonance. Identification of new bands.

Table 6.2 (continued).

Molecule	Spectral range, nm	Reference	Note
H_2	573–610	[6.38]	DC-excited discharge. Identification of new transitions between the $2p^3\Pi$ states and states in the vicinity of the dissociation limit.
I_2	520–630	[6.40]	DC-excited small-current discharge. $B \rightarrow X$ bands. Doppler-limited spatial resolution.
I_2	575–610	[6.41]	Medium-resolution spectra.
I_2	532	[6.42, 6.43]	Diode. Determination of the electron affinity energy.
CO	640–660	[6.44]	$D^3\Pi \rightarrow A^3\Pi$ bands.
CO	550	[6.45]	Vibrational-rotational transitions in the active medium of a CO laser.
NH_2	570–615	[6.38]	DC-excited discharge. First observations of free radicals.
NH_2	596–605	[6.24]	DC-excited discharge. Doppler-limited resolution.
NH_2	580–610	[6.39]	High-frequency discharge.
NH_2	570–600	[6.38, 6.46] [6.39]	DC-excited discharge. High-frequency discharge.
NO_2	6200	[6.47]	Diode IR laser. DC-excited discharge.
HCO	580–620	[6.26]	High-frequency CH_3CO discharge. Predissociation rate investigations.
CN	643–682	[6.24]	$A \rightarrow X$ bands.
NH_3	9500	[6.29]	DC-excited discharge, hollow-cathode discharge, high-frequency discharge. N_2O laser. Doppler-free resolution. Optical-microwave resonance.
NH_3	9500	[6.47]	Diode IR laser. DC-excited discharge.
NH_3	580–630	[6.39]	High-frequency discharge.
NH_3	9200–11250	[6.48, 6.49]	High-frequency discharge. Discretely tunable CO_2 laser.

Table 6.2 (continued).

Molecule	Spectral range, nm	Reference	Note
He_2	585–588	[6.24]	Rydberg transitions.
CO_2	9300–10800	[50–54]	Vibrational-rotational transitions in the active medium of a CO_2 laser.
D_2O H_2CO SO_2 H_2S H_2O	9100–9700	[6.48, 6.49]	High-frequency discharge. Discretely tunable CO_2 laser. Identification of transitions, search for transitions suitable for optically pumped far-IR lasers.
SF_6 N_2O	10600	[6.48, 6.49]	High-frequency discharge. Discretely tunable CO_2 laser Identification of transitions, search for transitions suitable for optically pumped far-IR lasers.
N_2^+	390	[6.55]	(0, 0) bands of the $x^2\Sigma \to B^2\Sigma$ transition. First observations of the optogalvanic spectra of molecular ions.
CO^+	490	[6.55]	(0, 0) band of the $x^2\Sigma \to B^2\Sigma$ transition.
NO	270–317	[6.56]	Flame. Multiphoton ionization.
PO	302–334	[6.56]	Flame. Multiphoton ionization.

As in the case of atomic spectroscopy, the advantages of the optogalvanic detection of molecular spectra become apparent on transitions of low optical density. We will restrict ourselves here to a few examples.

Typical research can be exemplified by the studies [6.26] into the absorption band structure of the HCO radicals on transitions terminating in a predissociation state. The linewidths provide information on the predissociation probability. The optical absorption technique cannot be applied here because density of the radicals is low. According to the estimates made by the authors of [6.26], the sensitivity of the optogalvanic technique is about 10^5 times that of its optical absorption counterpart, all things being equal. Optoacoustic detection is also difficult to implement because of the low density of the radicals and discharge noise.

Optogalvanic techniques made it possible to identify the bands of the systems $b^1\Sigma_g^+ \to a''^1\Sigma_g^+$ [6.27] and $c'_5{}^1\Sigma_u^+ \to a''^1\Sigma_g^+$ [6.28] and obtain Doppler-free spectra of the Ledbetter bands $c'_4{}^1\Pi \to a''^1\Sigma_g^+$ for the first time.

Examples of optogalvanic spectra obtained in the IR region by means of tunable diode lasers can be found in Fig. 2.20. To record NH_3 spectra, use was made in [6.29] of an N_2O laser tunable in the above-threshold amplification region of the desired vibrational-rotational transition. The high lasing power (around 1.5. W) allowed them to obtain Doppler-free NH_3 line profiles under saturated absorption conditions with relative ease. The authors of [6.30] used the double optical-microwave resonance technique, with the microwave frequency corresponding to the transition between the inversion doubling states. But the change in the magnitude of the optogalvanic effect, achieved on account of resonance when varying the microwave frequency was small because of the efficient collisional energy exchange between the levels even at low gas densities.

6.3 DETECTION OF RADIATION

The first proposal to use a discharge cell as a resonance radiation detector was made by the authors of [6.57] and experimentally verified using as an example the IR line of helium ($\lambda = 2058.1$ nm) in. They also suggested optogalvanic detection of Hg lines. More recently, it was shown in [6.58] that the spectral selectivity of such a detection technique is fairly high (around 0.1 cm^{-1}). If, for example, the detector is a hollow cathode filled with ^4He, the radiation from a ^3He-filled lamp cannot practically be detected. To make use of the high signal/noise ratio typical of the optogalvanic effect in conjunction with spectral selectivity, the authors of [6.59–6.62] applied a gas-discharge cell to detect the absorption of radiation in the intracavity laser spectroscopy technique. The experiments showed that such a detector provides for a gain of around an order of magnitude in the detection limit in comparison with the photoelectric multiplier.

Investigated as a CO_2 laser radiation detector in [6.49] was a high-frequency discharge in CO_2, N_2O, SF_6, and NH_3. The authors of this work noted in particular that the CO_2 lasing spectrum could be optogalvanically detected without any distortion if the detector used was a CO_2-filled high-frequency discharge cell.

6.4 LASER FREQUENCY STABILIZATION. WAVELENGTH CALIBRATION

In the case of active stabilization, the lasing frequency of a laser is controlled with reference to a certain datum. The feedback signal for the automatic frequency control (AFC) system is fed from a photodetector. In continuous-wave gas lasers, the lasing frequency is usually controlled with reference to the center of the amplification profile of the lasing transition itself. The error signal for the AFC system is produced by modulating the length of the laser cavity. The optogalvanic technique allows one to do away with photodetectors, the active medium of the laser itself being used as a detector.

Such a system, first implemented in [6.51] with a view to stabilizing the lasing frequency of a gas-discharge CO_2 laser with reference to the amplification profile center, is shown in Fig. 6.2. The laser consists of an active element – the discharge tube 1 supplied from the high-voltage source 2 – and a cavity formed by the mirrors $M1$ and $M2$. The mirror $M1$ is semitransparent and serves to couple the laser radiation out of the cavity. The mirror $M2$ is mounted on the piezoelectric corrector PC, and the cavity length modulation is effected by feeding the latter with the search a-c voltage (typically a few volts in amplitude and 1–10 kHz in frequency) produced by the generator 3. The optogalvanic signal is picked off the resistor 4 and is supplied to the input of the phase-sensitive detector 5, the search voltage serving as the reference signal. When the lasing frequency becomes detuned from the amplification profile center, the phase-sensitive detector produces an error signal which is amplified by the amplifier 6 and fed to the piezoelectric corrector.

If the lasing frequency modulated by the search voltage is, on the average, at the center of the amplification profile, the error signal is zero. Note that use can also be made of an electrodynamic mirror displacement element [6.52] instead of the piezoelectric corrector. In conditions of the experiment [6.51], the characteristic discharge voltage variations associated with the search signal amounted to some 10 V, or around 0.25% of the total voltage applied across the discharge tube. Smith and Moffat [6.53] described an improved electronic circuit for stabilizing the lasing frequency of a CO_2 laser with the aid of the optogalvanic effect. The

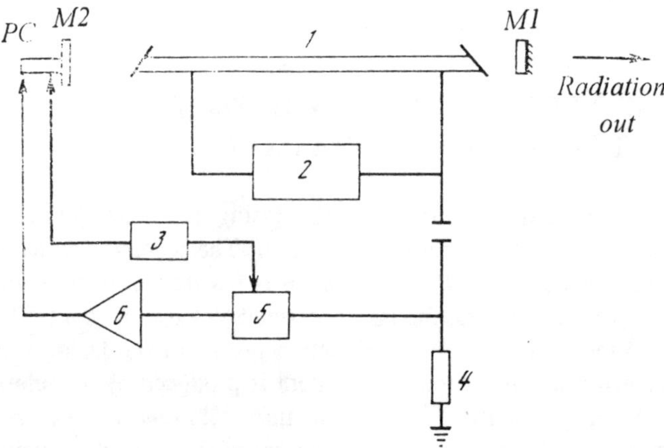

Fig. 6.2. Schematic diagram of optogalvanic frequency stabilization of a gas-discharge laser.

optimization of the conditions of [6.50] enabled the authors to realize a long-term frequency stability of around 5×10^{-9}. In [4.50–6.53], lasers with low active-medium gas pressure values (10–30 Torr) were studied, but the authors of [6.50] experimented with a waveguide CO_2 laser with an active medium pressure of 180 Torr. The sensitivity of the optogalvanic technique used to stabilize the frequency of the laser proved high enough to provide for a frequency instability of about 1 MHz (relative instability $\sim 10^{-8}$). Judging by the results obtained by various authors, the use of the optogalvanic effect provides for frequency instabilities of the same order of magnitude as those in photodetectors. Information on the frequency stabilization of He-Ne lasers by means of the optogalvanic effect can be found, for example, in [6.63, 6.64].

Optogalvanic laser frequency stabilization techniques are used not only in laboratory lasers, but also in their commercial counterparts, the Russian Model ЛГ-74 CO_2 laser included. With such a stabilization scheme, the amplification profile proper of the lasing transition plays the part of a frequency discriminator. To peak the discriminator in lasers with inhomogeneously broadened transitions, use can be made of the Lamb dip, or of the inverted dip, in a resonantly absorbing cell in the cavity. For the CO laser, such a technique was implemented in [6.66]. Use can also be made of an extracavity discharge cell [6.67]. Such cells are employed to stabilize and measure the lasing frequency in wide-band dye lasers [6.68–6.71]. The search signal is produced by varying the tilt angle of an intracavity etalon, and the error is produced as usual by a synchronous detector. The signal at the input of the synchronous detector typically ranges between 5 and 100 mV at a lasing power of 0.1–0.2 W. Stabilization occurs at discrete frequencies within the laser tuning range. In [6.68], use was made of the lines of neon (588.2 nm), sodium (589 and 616.1 nm), and barium (553.5 nm). The wavelength instability was in that case no more than 10^{-5} (the relative instability $\sim 2\times10^{-8}$). The number of resonance absorption lines in hollow-cathode discharges being great, this approach holds much promise for dye lasers. It was noted that stabilization could be effected even on lines with a very weak absorption.

The authors of [6.72–6.74] discussed the problem of the spectral calibration of the lasing line the determination of its width in tunable lasers. Hollow-cathode discharges with uranium lines in the IR, visible, UV regions of the spectrum are convenient to use for the optogalvanic stabilization and measurement of the laser wavelength [6.75]. An automatic system for the calibration of the wavelength of a laser with reference to 25 lines in the range 580–800 nm by means of a neon lamp was described in [6.76].

The usefulness of laser wavelength calibration with reference to the optogalvanic spectra of radicals in a discharge, specifically for laser-monitoring the composition of atmospheric species, was noted in [6.77].

One possible way to further develop optogalvanic laser-frequency stabilization and calibration techniques is to use extracavity discharge cells in a magnetic field. This will make it possible to displace the frequency datum in a controllable fashion and produce the error signal for the automatic laser frequency control system without any parasitic modulation of the laser radiation being caused by the position variation of the laser cavity elements.

6.5 ELEMENTARY PROCESS STUDIES

As noted earlier, the discovery of the optogalvanic effect in a discharge was associated with studies into the Penning ionization processes and the processes like $A^* + A \rightarrow A_2^+$ and $A^* + A \rightarrow A^+ + A^-$. These studies have been continued until very recently (see, for example, [6.78]).

Using a method essentially close to optogalvanic spectroscopy, the authors of [6.79–6.83] have investigated elementary processes occurring in electronegative gases. The electric signal being detected is produced under the effect of a CO_2 laser, or as a result of the Raman scattering of radiation from a solid-state laser and a dye laser, and is explained by the fact that the electron dissociative capture rate depends on the degree of vibrational excitation of the molecule of interest. Subject to studies were the SF_6, CCl_2F_2, and I_2 molecules. In [6.83, 6.84], the formation of ions in collisions between optically excited $Na(4d)$ atoms and O_2, SF_6, CH_3Br, and CCl_2F_2 molecules and the rate constants of the reactions involved were investigated.. Analysis of the data presented in [6.84] indicates that the Landau-Zener theory of nonadiabatic transitions in the system of molecular terms is applicable here.

The optogalvanic technique proves effective in the studies of two-photon dissociation processes [6.32–6.36]. The molecules are first raised to a repulsive or a predissociative state and then to a Rydberg state whence they get ionized upon thermal collisions. And even if the molecule has a wide excitation band, the resultant optogalvanic effect will be of resonance character (hybrid resonances). Specifically this makes it possible to study the selectivity of the population of atomic states in dissociation.

The optogalvanic effect was used in [6.85] to investigate absorption in laser-induced dipole-dipole interaction collisions leading to the redistribution of the level populations of particles:

$$Ba(6s^2\,{}^1S_0) + Ba(6s^2\,{}^1S_0) + h\nu(339.4\ \text{nm}) \rightarrow Ba(6p^1P_1) + Ba(5d^1D_2),$$

$$Ba(6s^2\,{}^1S_0) + He(2s^1S_0) + h\nu(615.1\ \text{nm}) \rightarrow He(2p^1P_0) + Ba(5d^1D_2),$$

the latter reaction being observed to occur for the first time. Similar processes involving the formation of ions in light-induced collisions between alkali metal atoms were investigated in [6.86]. It is anticipated that this method will also be used to study multipole interactions, multiple-photon processes, and interactions involving charge exchange upon collisions in light fields.

Using a combination of optogalvanic signal detection and fluorescence techniques, the authors of [6.87] investigated the mechanisms of the singlet-triplet conversion of atoms in a neon plasma.

Doppler-free spectroscopy makes it possible to study the impact broadening and shifts of spectral lines to a high degree of precision (see, for example, [6.88]). The collision ionization probability of Rydberg atoms was measured in [6.89]. One may also recall optogalvanic effect investigations on reactions involving particles with oriented spin magnetic moments [6.90], the vibrational relaxation of molecules [6.91], collision relaxation rate of excited states of helium [6.92–6.94], lifetime of highly excited states of neon [6.95], affinity energy in the photodetachment of electrons [6.96], and so on.

6.6 QUANTITATIVE PLASMA SPECTROSCOPY AND DIAGNOSTICS

Research has shown that optical absorption and optogalvanic spectra are, generally speaking, not similar (see, for example, Fig. 2b and [6.97]). This is because the optogalvanic effect depends not only on the amount of light absorbed, but also on the mechanism responsible for the change in the conductivity of the object of study, which involves the upper level of the transition in hand. Based on these qualitative considerations, one can expect that the similarity between optical and optogalvanic line spectra should be observed in the case of transitions with a common upper level, or where the upper levels of transitions exist in closely similar conditions as regards ionization. The latter condition is met, for example, by molecular spectra with their fine rotational structure. The system of molecular rotational levels with widely ranging rotational quantum numbers is, as a rule, very compact and gets mixed effectively by collisions, which is used in traditional spectroscopy to determine the temperature of gases. These possibilities have as yet not been systematically studied. Note the measurements taken in [6.55] of the intensity of lines in the rotational structure of the optogalvanic spectra of $N_2^+ (B^2\Sigma, v' = 0 \rightarrow X^2\Sigma, v'' = 0)$ in a constricted discharge. The rotational line intensities were observed to alternate in the ratio 2:1 in accordance with the nuclear spin. The set of lines for ions of the same nuclear spin orientation had their intensities correspond to a Boltzmann distribution among the rotational levels, and the optogalvanic signals were proportional to the radiation intensities used. This definitely points to the possibility of quantitative investigations into nonequilibrium plasmas on the basis of their optogalvanic line spectra.

The study of line profiles is an individual division of quantitative plasma spectroscopy and diagnostics. Such studies were already conducted in the very first publications on optogalvanic spectroscopy with the aid of lasers [6.98, 6.99]. As noted in Sect. 2.1, in the simple model of a two-level atom under weak absorption conditions in a high-frequency discharge a similarity between optical and optogalvanic line profiles exists. This fact was experimentally verified in a number of works, and is now being used to determine, among other things, the temperature of gases from the

Doppler broadening of their optogalvanic spectral lines (see, for example, [6.100, 6.101]). Of interest are the investigations performed in [6.55] on the Doppler-broadened optogalvanic line profiles of molecular ions having a nonisotropic velocity distribution. Such works might result in determining particle velocity fields in various objects through the solution of an inverse problem wherein the given data is a Doppler-broadened profile of generally non-Gaussian shape, as in the case of emission spectroscopy [6.102].

At the same time, there are data [6.97] pointing to substantial distortions of optogalvanic line profiles. Figure 6.3 shows neon line profiles (λ = 585.2 nm) recorded at various hollow-cathode discharge currents. At currents below 100 mA the line profiles recorded by both the absorption and the optogalvanic technique coincide (the asymmetry

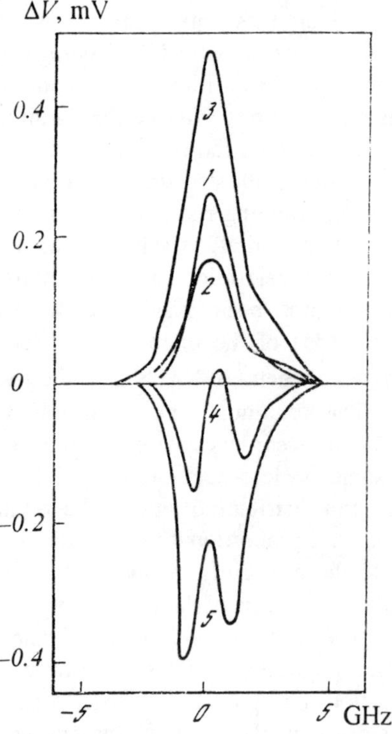

Figure 6.3. Optogalvanic line profiles of neon (transition $1s_2 \to 2p_1$, λ = 585.2 nm) in a hollow-cathode discharge [6.97]. Gas pressure 3 Torr; discharge current mA: *1* − 20; *2* − 100; *3* − 300; *4* − 400; *5* − 450.

observed at small currents is due to the isotope shift). As the discharge current is increased, the shape of the optogalvanic line profile changes drastically. The signal polarity changes and a structure appears in the profile. Similar effects were also observed in the IR region of the spectrum [6.47, 6.103]. It was noted in [6.97] that distortions took place when the absorption of light by the plasma was high ($\geq$ 20%). Where the optical density of the plasma is high, radiation at the line-wing frequencies effectively interacts with it all over the beam path, whereas radiation at the center frequency of the transition gives rise to an optogalvanic effect with a maximum in the beam entrance zone. In an inhomogeneous plasma, the resultant line profile cannot simply be obtained by averaging the optogalvanic signal over the various discharge zones. But where the optical density of the plasma is low, the agreement between the absorption and optogalvanic detection results is good. Figure 6.4 presents the neon line profiles recorded in [6.97]. The case illustrated in Fig. 6.4a corresponds to a low absorption (less than 3%) on the transition $1s_5 \rightarrow 2p_2$. The hollow-cathode discharge current is 50 mA. The case illustrated by Fig. 6.4b corresponds to absorption on the Zeeman components of the transition $1s_5 \rightarrow 2p_4$ in an indicator lamp. The magnetic field strength is 1600 Gs, discharge current, 10 mA, and line-center absorption, 12%. The agreement between the absorption and optogalvanic line profiles is satisfactory. The slight differences are apparently due to the low signal/noise ratio in the absorption measurements.

Thus, experience shows that optogalvanic spectroscopy can frequently be successfully used for quantitative measurement and plasma diagnostics purposes. The applications associated with the recording of spectral line profiles in low optical density conditions appear to be substantiated the most at this stage.

One low-temperature plasma diagnostics problems which requires further work is the determination of the negative ion densities. Proposals exist to use the optogalvanic effect for this purpose. The authors of [6.104] investigated a dc-excited hollow cathode discharge in NF_3 at a pressure of 0.3 Torr. The discharge was irradiated by a pulsed Xe-Cl laser beam (λ = 308 nm), which caused electron photodetachment and gave rise to a galvanic signal. The discharge current variations might be associated with variations in the electron density n_e, which could be reliably measured by the radio-frequency technique under the above experimental conditions. The change of the electron density under the effect of a light pulse is related to the negative ion density variation by the relation

$$\Delta n_e(t) + N^-(t) = N_0^-, \qquad (6.1)$$

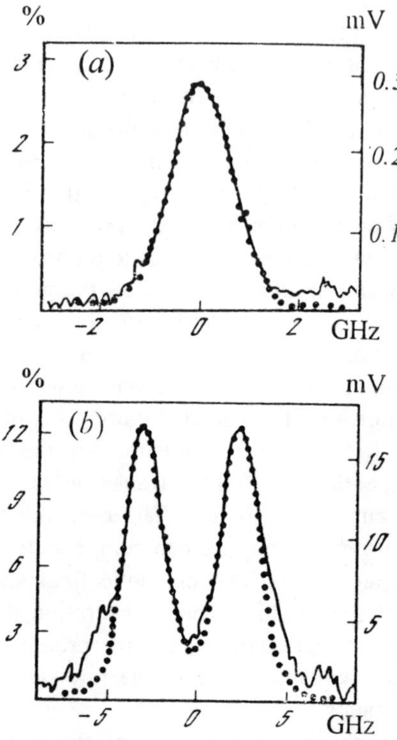

Figure 6.4. Neon line profiles recorded in [6.97]. The solid curves correspond to optical absorption measurements and the dotted ones, optogalvanic measurements, (*a*) Hollow-cathode discharge, $1s_5 \rightarrow 2p_2$ transition, $\lambda = 597.5$ nm, discharge current 50 mA, gas pressure 3 Torr; (*b*) neon indicator lamp, σ-components of the $1s_5 \rightarrow 2p_4$ transition, $\lambda = 594.4$ nm, magnetic field strength 1600 Gs.

where N_0^- is the steady-state negative ion density. If σ is the photodetachment cross section and I is the intensity of radiation of frequency ν, then

$$\frac{d\Delta n_e}{dt} = \frac{\sigma I}{h\nu}\left(N_0^- - \Delta n_e\right), \tag{6.2}$$

$$\Delta n_e(t) = N_0^-\left[1 - \exp\left(-\frac{\sigma}{h\nu}\int I\,dt\right)\right]. \tag{6.3}$$

Qualitative mass-spectrometric analysis revealed that the principal negative ion was F^-, and it was for this ion that the cross section σ was taken.

The experimental relationships between $\Delta n_e(t)$ and the laser intensity were approximated by formula (6.3) to determine N_0^-. Figure 6.5 presents a similar relationship borrowed from [6.104]. In this work, the density N_0^- $= 2.4{\times}10^9$ cm^{-3}, which is approximately 50 times the electron density.

The optogalvanic effect in high-frequency discharges in the electronegative gases Cl_2 and Bcl^3 was studied in [6.105] at a high-frequency field frequency of 50–150 kHz, and also in Cl_2 at a frequency of 13.56 Mhz in [6.106]. The measurements taken in these works also yielded $N_0^-/n_e \sim 50$. The ions O^- were detected in [6.107, 6.108].

The optogalvanic effect was used in [6.109] to take local electric field strength measurements in a neon discharge plasma. The field strength was found from the Stark splitting of the states for which the linear Stark effect was great because of the closeness of levels differing in parity. On the other hand, the splitting could reliably be calculated as a function of the field strength. The measurement region volume was localized to some 10^{-3} cm^{-3} by means of the two-step excitation scheme using intersecting laser beams already discussed above. Use was made of two dye lasers pumped by a pulsed nitrogen laser. One of the lasers generated 588.2-nm light pulses

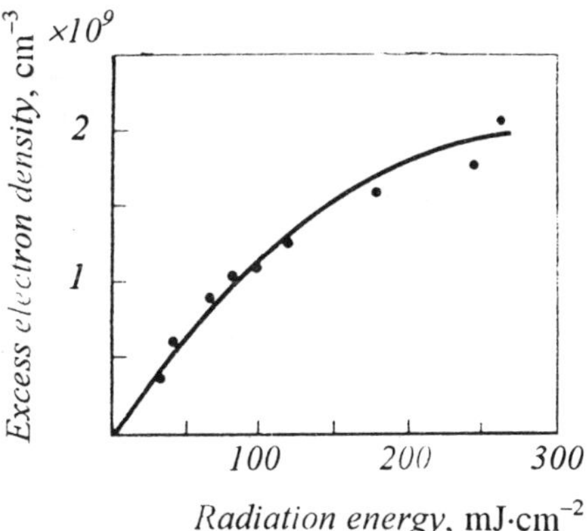

Figure 6.5. Excess electron density as a function of the laser intensity in electron photodetachment [6.104]. NF$_3$ discharge. The dots represent experimental data points and the solid curve – theory. $N_0^- = 2.4{\times}10^9$ cm^{-3}.

that raised the Ne atom from the metastable state $2p^5 3s^2 P_2$ to the state $2p^5 3p^3 P_1$ and the other operated after a delay of a few nanoseconds to produce 439.3-nm pulses that excited the $2p^5 3p^3 P_1$ atom to move to the Stark multiplet $2p^5(2P_{1/2})11d'$. The delay between the laser pulses was necessary to eliminate the optogalvanic effect at the first excitation step. The choice of the excitation scheme with the intermediate level $2p^5 3p^3 P_1$ was due in particular to the fact that laser excitation at the first stage produces no strong optogalvanic effect, the associative ionization channel being inoperable because of the low energy of state. On the contrary, the second step provides for effective ionization precisely by this channel.

The advantage of the optogalvanic detection technique in comparison with optical absorption measurements, as in many other cases, is a high sensitivity and good signal/noise ratio, which is of principal importance in conditions of sharp spatial localization.

The experiments in [6.109] were conducted with a discharge between plane aluminum electrodes 1.5 cm in diameter spaced 0.4 cm apart. The neon pressure was 2 Torr.

Figure 6.6 presents exemplary optogalvanic Stark component spectra recorded in [6.109] at various points in the discharge between the anode and the cathode while varying the lasing frequency of the second-step laser, and Fig. 6.7 shows the variation of the electric field strength in this experiment. Such investigations are of particular interest in determining the drift rates and spatial distributions of ions, calculating ion currents, analyzing ionization mechanisms, etc.

Similar studies of a helium discharge were carried out in [6.110].

To date, measurements of electric fields in discharges by the optogalvanic technique have been developed considerably. To illustrate, the use of high-lying Rydberg states for which the linear Stark effect is great provides for a sensitivity of the order of 1 V/cm [6.111]. Illuminating the plasma with a variable-polarization radiation makes it possible to measure not only the strength of the field, but also the direction of its vector [6.112]. Field measurements can also be taken while illuminating the plasma on molecular spectral lines [6.113]. Some results of studies on electric field measurements in glow discharge plasmas by the optogalvanic technique have been described in more detail in the review [6.114].

Note also that the authors of [6.115] hold that the time dependence of the optogalvanic signal consists of two components, one of them being due to ionization processes and the other, to optoacoustic ones. The conclusion is drawn on this basis that the optogalvanic signal can be used as a high-sensitivity probe for studying various processes in plasmas, such as the mobility of electrons, radiation trapping, excitation transfer among atoms,

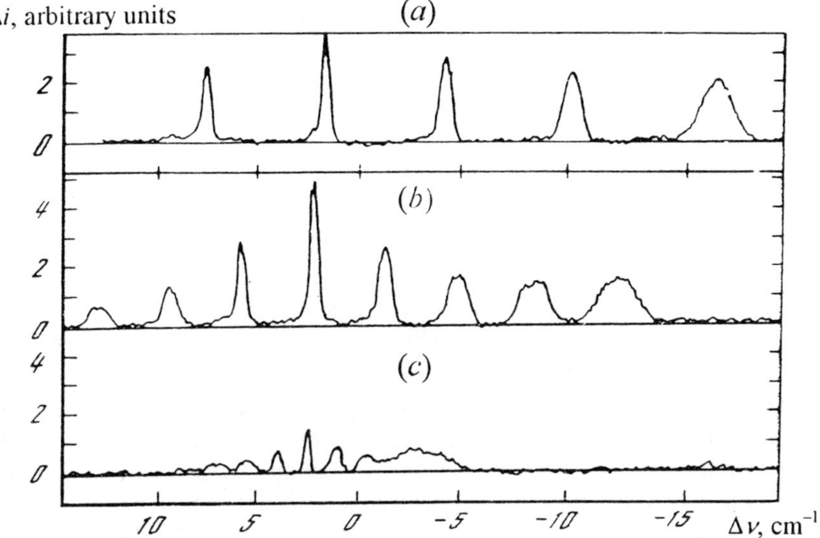

Figure 6.6. Optogalvanic Stark component spectra recorded at various distances from the cathode [6.109]. The optogalvanic signal intensity Δi is plotted in arbitrary units. The zero frequency detuning corresponds to the frequency of the transition $2p^53p^3P_1 \rightarrow 2p^5(^2P_{1/2})11d'$ in the absence of the field. (a) the distance from the cathode, $l = 0.025$ cm, electric field strength $E = 3958$ V/cm; (b) $l = 0.152$ cm, $E = 2420$ V/cm; (c) $l = 0.254$ cm, $E = 1044$ V/cm.

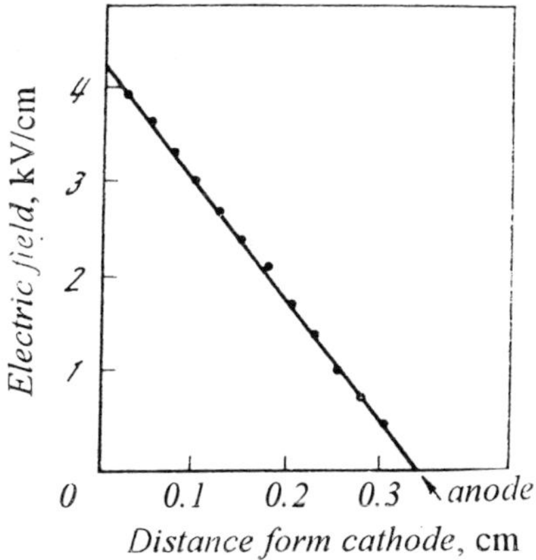

Figure 6.7. Spatial electric field distribution in the cathode region in [6.109].

and so on. It is expected that a low-pressure radio-frequency discharge will be used for the purpose.

6.7 COMPOSITION ANALYSIS

Green and co-workers [6.116] were the first to suggest using the optogalvanic effect for analytical purposes. Considering the high sensitivity of the technique, this application seems quite natural.

The relationship between the capabilities of the optogalvanic technique and those of absorption spectroscopy or other optical techniques, including laser ones (laser-induced fluorescence, optoacoustic detection, intracavity spectroscopy), Raman scattering or electron paramagnetic resonance techniques selective in the energy states of atoms or molecules can hardly be universal. It depends on the optogalvanic effect mechanism, the type of the transition involved and the object wherein it manifests itself, and the scheme used to implement it. It is therefore advisable at this stage to orient oneself toward the results already achieved.

The experimental studies conducted in [6.117] with hollow-cathode discharges in Na, U, Eu, and Zr showed that the current fluctuations in the supply circuit of the specially designed hollow cathode corresponded to statistical noise. This established the detection capability of the optogalvanic detection technique at some 10^6 cm^{-3} (or around 10^2 cm^{-3} in the case of metastable states), which was practically attained. The authors of [6.118] observed the optogalvanic effect and recorded the absorption spectra of the Ba$^+$ and Eu$^+$ ions in a hollow-cathode discharge. The ions U$^+$ were detected in [6.13], I$^+$ in [6.40], Sr$^+$ in [6.11], Pr$^+$ in [6.119], and N$_2^+$ and CO$^+$ in [6.55].

The majority of the works on the optogalvanic analysis were performed using flames, the results being easy to obtain and reproducible. The well-developed methods for introducing the substance of interest in the flame enable one to carry out reliable calibrations and practically remove the above-discussed difficulties associated with the taking of quantitative measurements in nonequilibrium objects. Table 6.3 lists the optogalvanic detection limits for a number of elements, along with the detection limits attained with other methods. The data are mainly borrowed from [6.120–6.125]. The following designations are adopted for the various techniques: OG – optogalvanic, AB – absorption, EM – emission, Fl$_I$ – fluorescence by incoherent irradiation, and FL$_L$ – fluorescence by laser irradiation.

The typical detection limits range between 10^{-3} and 10^2 ng/ml. For a

Table 6.3. Detection limits in flames [6.120–6.125].

Element	Wavelength OG, nm	OG	AB	EM	FL_I	FL_L
				Detection limit, ng/ml		
Ag	328.1	1	1	2	0.1	4
Al	309.3	0.1				
Au	242.8	1				
Ba	307.2	0.2	20	1		8
Bi	306.8	2	50	20000	5	3
Ca	300.7	0.1	1	0.1	20	0.08
Cd	228.8	0.1				
Co	252.1	0.08				
Cr	298.6	2	2	2	5	1
Cr	301.8	2				
Cs	455.5	0.004				
Cu	282.4	100	1	0.1	0.5	1
Cu	324.8	100				
Fe	298.4	4				
Fe	302.1	2	4	5	8	30
Ga	287.4	0.07	50	10	10	0.9
Ga	294.4	0.1				
In	303.9	0.006	30	0.4	100	0.2
K	404.4	0.1				
Li	610.4	0.012				
Li	639.3	0.4				
Li	670.8	0.001	1	0.02		0.5
Lu	308.2	0.2	1	0.02		0.5
Mg	285.2	0.1	0.1	5	0.1	0.2
Mn	279.5	0.02	0.8	1	1	0.4
Na	285.3	0.05	0.8	0.1		0.1
Na	589	0.01				
Ni	300.2	0.08	5	20	3	2
Pb	280.2	0.6	10	100	10	13
Pb	283.3	0.09				
Rb	420.2	0.1				
Sc	301.9	0.2				
Sn	284	0.3				
Sn	286.3	2	20	100	50	
Sr	460.7	0.4				
Tl	291.8	0.09	20	20	8	4

number of elements, such as Pb, transitions between excited states provide for lower detection limits than transitions from the ground state. In many cases, the detection limits attained with the optogalvanic technique are perceptibly lower than those typical of the other techniques, the linearity in concentration ranging between 10^4–10^5 and more. Estimates made in [6.123] show that the electric-noise-dominated detection limit in flames amounts to around 10^5 cm^{-3}. This limit was actually reached for Li.

The development of this analysis technique proceeds toward the optimization of the excitation and detection schemes, as well as the object geometries. The influence of the electrode configuration and the gas composition and flow rate are being studied, and various detection methods (overall voltage drop, probe potential, microwave absorption) are being compared. The authors of [6.124] developed an analytical instrument with a dynamic range of no lower than 10^4 and capable of analyzing elements in the range 217–700 nm. Certain prospects might be associated with the introduction of inductively coupled plasma sources already noted in Ch. 3 into analytical practice. For example, the optogalvanic detection technique was used in [6.126] to measure the $^{13}CO_2/^{12}CO_2$ ratio in the air exhaled by man. In that case, the heavy isotope was used as a label. The concentration ratio was determined from the change in the impedance of low-power discharges illuminated with an amplitude-modulated radiation from the appropriate isotope-replaced CO_2 lasers.

6.8 COOLING OF GAS-DISCHARGE PLASMA BY EXTERNAL RADIATION

The absorption of a resonance radiation in a gas-discharge plasma may cause its temperature and particle concentration to change. The optogalvanic variation of the temperature of neutral particles is especially pronounced in the excitation of molecular discharges [6.127]. Irradiation first of all causes changes in the concentration and temperature of electrons. Later on, the altered energy balance causes the temperature of the neutral particles to vary as well.

Arimondo and co-workers [6.128, 6.129] used an optoacoustic technique to measure gas temperature variations consequent upon the optogalvanic effect. The experiments were conducted in the positive column of a He-Ne discharge 21.5 cm long and 0.25 cm in radius at a gas pressure of 0.4–4 Torr and a discharge current of 1–5 mA. The radiation used to illuminate the discharge along its axis was modulated with a frequency of 310 Hz. The acoustic waves resulting from the gas temperature variations were detected by a microphone mounted on the inside wall of the discharge tube. Fig. 6.8 presents the optogalvanic and optoacoustic spectra due to the transition at $\lambda = 587.6$ nm in helium and a number of transitions in neon. As can be seen, irradiation can cause both the heating and cooling of the gas. And while the heating of the gas under the effect of laser radiation is easy to understand from the energy balance standpoint, the gas cooling effect is not obvious. Essentially, the matter is as follows [6.30].

The plasma contains atoms having both short-lived and metastable excited states. The latter play an important part in the gas ionization processes and hence in the formation of the plasma itself. If an atom acquires an energy of $\hbar\omega_1$ from an external resonant source and rises as a result from the metastable level A to the state B (see Fig. 6.9) which is optically coupled to the level C coupled in turn with the ground state D, the spontaneous decay of the levels B and C results in the emission of an energy $\hbar\omega = \hbar\omega_2 + \hbar\omega_3$, such that $\hbar\omega > \hbar\omega_1$, i. e., the atom cools down.

Such a plasma cooling scheme is easy to implement, for example, in neon. As already stated, the block of the lower excited states $1s$ of this atom consists of two metastable levels ($1s_3$ and $1s_5$) and two resonance levels ($1s_2$ and $1s_4$). The resonance levels live shorter than their metastable counterparts, even at low and medium gas pressures. The states in the upper block $2p$ are optically coupled to the levels $1s_i$. Therefore, following the excitation of the transition $1s_5 \rightarrow 2p_4$, for example, and subsequent

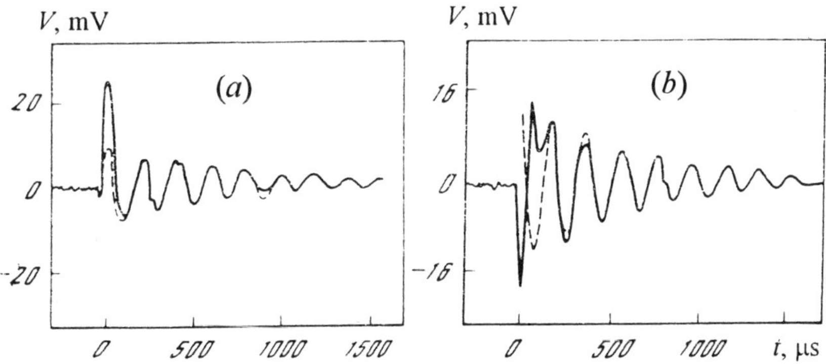

Figure 6.8. (a) Optogalvanic and (b) optoacoustic signals in a He-Ne discharge (Ne – 1 Torr, He – 1.6 Torr) at a current of $i = 2.5$ mA [6.129].

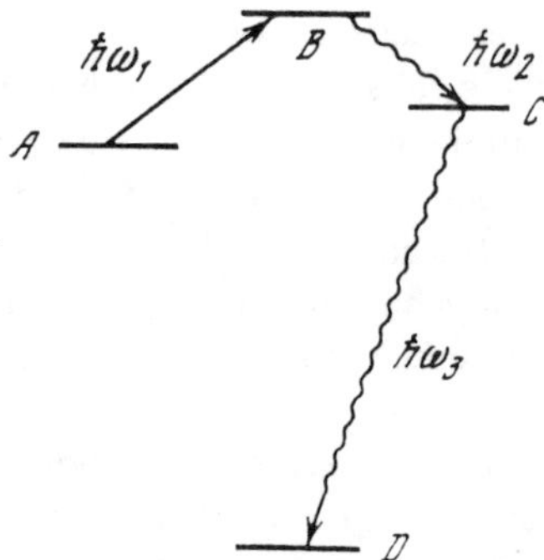

Figure 6.9. Radiative plasma cooling model.

decay of the level $2p_4$, the levels $1s_2$, $1s_3$, and $1s_4$ get populated more. The spontaneous decay of the resonance states $1s_2$ and $1s_4$ depopulates the lower block $2p$. This reduces the gas ionization rate and, accordingly, the electron concentration n_e and the discharge current i. If the initial voltage drop across the discharge proves insufficient to sustain it in the new conditions, the cooling of the gas can result in the quenching of the gas discharge.

To estimate the efficiency of the cooling process, use can be made of the optogalvanic effect model described earlier (see Sect. 2.4). Figure 6.10 illustrates the energy density variation of the positive column plasma:

$$W = \sum n_i E_i + \left(3/2\right) n_e k T_e . \tag{6.4}$$

Here E_i is the excitation potential of the ith level. The estimate was made on the assumption that the pulsed radiation saturated the resonance transition at the initial instant of time. With the discharge current in the absence of irradiation, $i(0)$, being equal to 10 mA (curve 1 of Fig. 6.10), the discharge continued to glow following its temporary cooling, but was extinguished at $i(0) = 0.13$ mA (curve 2).

This phenomenon was experimentally observed in a discharge with a positive column length of $L = 1$ cm at a discharge current of $i(0) = 0.13$ mA. The duration of the laser pulse was around 10 ns, its power, some 10 kW, linewidth, about 0.05 nm, and wavelength $\lambda = 594.48$ nm ($1s_5 \rightarrow 2p_4$ transition). The results of the experiment on the laser quenching of the

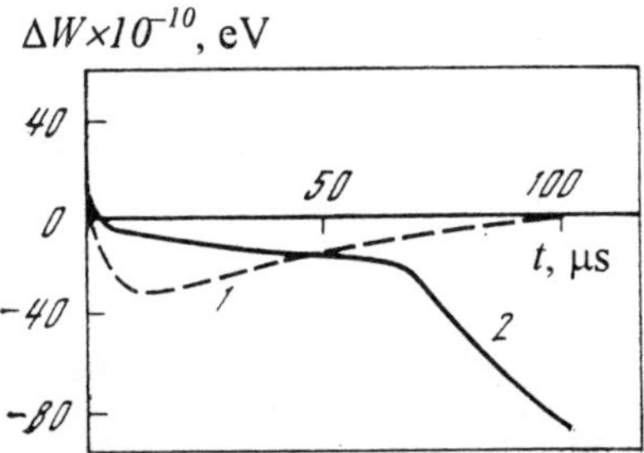

Figure 6.10. Light-induced variation of the energy of the gas-discharge plasma. *1 –* $i(0) = 10$ mA; *2 – $i(0) = 0.13$ mA.*

discharge are presented in Fig 6. 11. The arrow indicates the instant the laser was fired.

6.9 ISOTOPE SEPARATION

Bridges [6.131] suggested using the optogalvanic effect for isotope separation purposes. This technique is based on the fact that laser radiation excites a transition in one of the isotopes of the given element, which later the causes the concentration of the ions of this isotope to increase. The subsequent extraction of these ions enriches the isotope mixture.

The author of [6.132] theoretically analyzed the optogalvanic isotope enrichment of Cu ions in the positive column of Cu-Ne discharges in quasistationary laser excitation conditions on the basis of the results

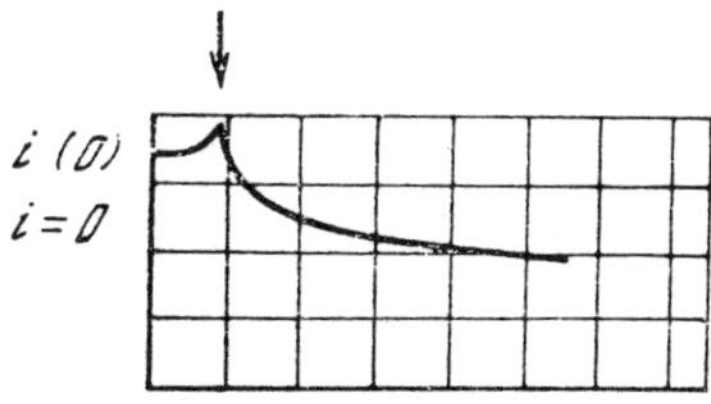

Figure 6.11. Light-induced quenching of a glow discharge. Scan – 200 µs per division.

obtained in solving balance equations. The model used allowed for electron temperature variation, elastic and superelastic electron collisions, the gas ionization and heating processes. Subject to analysis were isotopes 63 and 65 a. m. u. in mass. The process was analyzed at a constant discharge current for the case of excitation of the $^2D_{5/2} \rightarrow 2P_{3/2}$ transition ($\lambda = 510.6$ nm) in the isotope with a mass of 63 a. m. u. under laser saturation conditions. The separation magnitude depended on the initial isotope concentration, discharge current density, and discharge tube radius.

Included in the analysis were six levels of the copper atom and one level in neon, the neon atom being modeled to have three levels and a single ion state (Fig. 6.12).

In addition to the levels between which the transition occurred under the effect of laser radiation, the ground states of the copper atom and ion and also a quasilevel which modeled highly excited states of the copper atom were included in the analysis. The presence of the quasilevel enabled the author to take account of stepwise ionization processes. The neon gas played the role of a buffer gas stabilizing the discharge, the level structure of neon also being considered when calculating the kinetics of the process to determine in particular the electron temperature and allow for the Penning processes. Considered in the neon atom were not only the ground levels of the atom and ion, but also metastable states and a quasilevel, as in the case of the copper atom. The model describing the population kinetics included electronic excitation, quenching, and ionization, the Penning ionization, charge exchange, the diffusion of atoms and ions onto the dis-

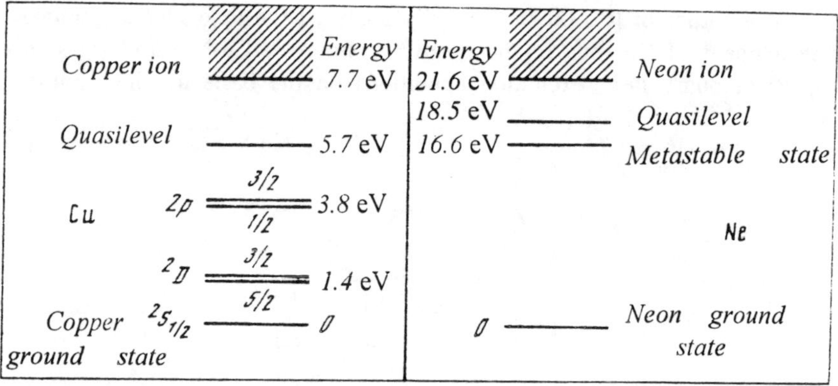

Figure 6.12. Atomic states of copper and neon considered in [6.132].

charge tube wall, radiative recombination, spontaneous emission and trapping of radiation, and, of course, the laser-induced resonance radiative processes.

The cross-section-averaged gas temperature $\overline{T}_g$, defined by the heating of the gas and the loss of heat on the discharge tube wall, was found by solving the heat conduction equation

$$\overline{T}_g = T_w + jER/2k_T, \tag{6.5}$$

where T_w is the wall temperature, j is the mean current density, and k_T is the thermal conductivity coefficient. The current density is given by

$$j = i/\pi R^2, \tag{6.6}$$

where the total discharge current is

$$i = 1.36 n_e R^2 \mu_e e E. \tag{6.7}$$

The enrichment factor is

$$\beta_i = f_{Ii}/f_{Ni} - 1, \tag{6.8}$$

where f_{Ii} is the ration between the ion densities of the two isotopic species and f_{Ni} is a similar ratio for the neutrals.

Figure 6.13a shows the density of the copper atoms in the lower excited state ^2D and the electron concentration in the absence of the resonance laser irradiation.

The variations of the parameters T_e, n_e, and ΔV under the effect of the resonance laser radiation are illustrated in Fig. 6.13b. The enrichment factor for the ions 63 a. m. u. in mass, β_{63}, under absorption saturation conditions is shown as a percentage in Fig. 6.14. The pressure of the neon gas is 15 Torr, that of the copper vapor, 0.1 Torr, $T_w = 1600°C$, and the discharge tube diameter is 1 cm.

The decisive reason for the rise of β with the increasing discharge current is the growth of the $^2D_{5/2}$ copper atoms.

Note that in the above system transport of matter due to the diffusion gas flow and ion flux (the cataphoresis phenomenon [6.133]) takes place. Let the gas consist of a mixture of two isotopes with densities $n^{(1)}$ and $n^{(2)}$ responsible for ionization processes and a buffer gas governing the

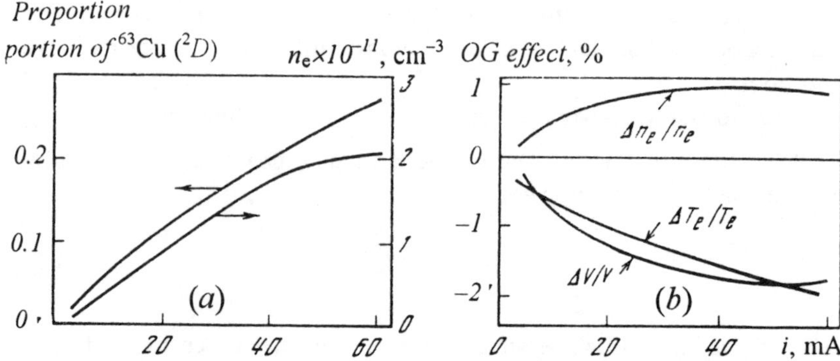

Figure 6.13. Typical optogalvanic effect parameters considered in [6.132] for the copper isotope with a mass of 63 a. m. u. in a concentration of 0.3. (a) No laser irradiation at λ = 510.6 nm; (b) illustrating changes in the electron density, electron temperature, and discharge voltage consequent upon resonance laser irradiation under transition saturation conditions.

transport process coefficients. The flux of ions of species k is

$$\Gamma_k^{(i)} = \frac{\sum\limits_{i=1}^{r} S_i n_i}{\sum\limits_{k=1,2}\sum\limits_{i=1}^{r} S_i n_i^{(k)}} \frac{\mu_i}{\mu_e} \frac{i}{e}. \tag{6.9}$$

Here μ_i and μ_e are the ion and electron concentrations, respectively, e is the electronic charge, S_i is the coefficient of electron impact ionization from the ith atomic state, and r is the number of levels considered.

The charged particle balance is formed by the electronic ionization of the atoms and the ambipolar diffusion drift of the particles onto the wall enclosing the cylindrical discharge space. In that case,

$$\sum\limits_{k=1,2}\sum\limits_{i=1}^{r} S_i n_i^{(k)} = (2.4/R)^2 D_e \mu_i / \mu_e, \tag{6.10}$$

where D_e is the electron diffusion coefficient. For simplicity, we will assume that the atoms $n^{(2)}$ are mainly ionized from the ground state, whereas their counterparts $n^{(1)}$, from an excited state, so that $n_2^{(1)} \cong \delta n_1^{(1)}$,

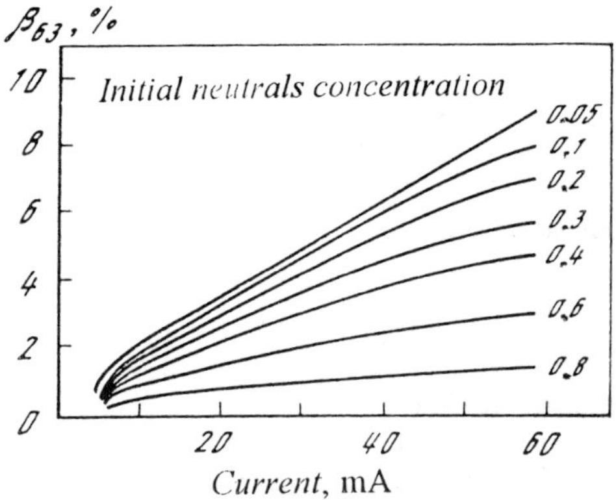

Figure 6.14. Ion enrichment factor β_{63} for the copper isotope 63 a. m. u. in mass as a function of the discharge current and initial concentration of this isotope. The discharge tube diameter is 1 cm [1.132].

where δ has the meaning of the degree of excitation saturation. In that case, we have from Eqs. (6.9) and (6.10)

$$\Gamma_k^{(i)} = \alpha_k n_1^{(1)}, \tag{6.11}$$

$$\alpha_1 = \frac{S_2 \delta i}{\left(2.4 / R\right)^2 D_e e}, \quad \alpha_2 = \frac{S_1 i}{\left(2.4 / R\right)^2 D_e e}. \tag{6.12}$$

Let us now write the transport equation for the ions of species k:

$$\frac{d}{dx}\left[-D_k \frac{dn_1^{(k)}}{dx} + \alpha_k n_1^{(k)}\right] = 0, \tag{6.13}$$

where $D_{1,2}$ is the atomic diffusion coefficient. Let the isotope vapor be formed in a vaporizer ($x = 0$) and deposited on a cooler ($x = L$) where the concentrations of both isotopes are specified. Solving equation (6.13), we get the following expression for the isotope flux onto the cooler:

$$\Gamma_k = \frac{\alpha_k n_1^{(k)}(L = 0)\exp\left(\alpha_k L / D_k\right) - n_1^{(k)}(L)}{\exp\left(\alpha_k L / D_k\right) - 1}. \tag{6.14}$$

We define the isotope separation factor as the ratio

$$Q = \frac{\Gamma_1}{\Gamma_2}\left[\frac{n_1^{(1)}(L=0)}{n_1^{(2)}(L=0)}\right]^{-1} \tag{6.15}$$

and consider various situations. Let

$$\alpha_k L/D_k \ll 1. \tag{6.16}$$

In that case, considering that

$$n_1^{(k)}(L=0) \gg n_1^{(k)}(L), \tag{6.17}$$

we get

$$\Gamma_k \cong n_1^{(k)}(L=0)/(L/D_k), \quad q \cong D_1/D_2, \tag{6.18}$$

i. e., predominant in this case are diffusion conditions and the isotope separation process goes but weakly because $D_1 \cong D_2$. When

$$\alpha_k L/D_k \gg 1, \tag{6.19}$$

the diffusion processes are unimportant and

$$q \cong \alpha_1/\alpha_2 \cong S_2\delta/S_1. \tag{6.20}$$

For

$$\alpha_1 L/D_1 \gg 1, \quad \alpha_2 L/D_2 \ll 1, \tag{6.21}$$

$$q \cong (\alpha_1 L/D_2)\exp(\alpha_1 L/D_1). \tag{6.22}$$

In that case, the second isotope is conveyed on account of diffusion processes, while the first one, on account of the ion current. At $E = 1$ V·cm^{-1}, $L = 10$ cm, and $T_g = 600$ K, $\alpha_1 L/D_2 \cong 20$ and $q \cong 4\times10^5$.

6.10 DETECTION OF PLASMA-CHEMICAL REACTION PRODUCTS

In the molecular plasma of a glow discharge complex processes causing the formation of chemical compounds, including radicals, may occur. Changes in the gas composition cause alterations in the electrical and optical characteristics of the plasma and can be detected by the optogalvanic technique.

For example, a characteristic plasma-chemical reaction occurring in a mixture of nitrogen and oxygen is the formation of nitrogen oxides. The formation of NO is an endothermal reaction and proceeds at high temperatures. The No formed may thereafter become further oxidized to form nitrogen dioxide NO_2. This reaction is, on the contrary, exothermic and easily takes place at ordinary temperatures. The equilibrium concentration of nitrogen monoxide, NO, at an air temperature of 2000 K amounts to 7×10^{-3} cm^{-3} and that of NO_2, 10^{-4} cm^{-3} [6.134]. In gas-discharge conditions, especially where the discharge plasma is of nonequilibrium character, the yield of plasma-chemical reaction products may be much higher because of the excitation of the molecules by electron impact [6.135].

The plasma-chemical process of formation of nitrogen oxides was observed in [6.136] in a discharge chamber whose design is shown schematically in Fig. 6.15. The discharge in the metal chamber was initiated between two copper electrodes which was cooled with water to improve the discharge stability and prevent the glow-to-arc transition. The discharge glowed stably in the current range 30–200 mA at a gas pressure from a few torrs to an atmosphere. The maximum discharge length in air at atmospheric pressure depended on the voltage of the power supply source, reaching 1.5 cm at 2 kV, the discharge current density amounting to 5–10 A/cm^2. The discharge length could be varied by moving the bellows-mounted electrodes. The discharge was laser-irradiated at right angles to the plasma column through the quartz windows in the discharge chamber walls. The discharge circuit included the power supply source E, capacitor C 0.15 μF in capacity serving to reduce the output impedance of the supply source, the movable electrodes with the discharge gap in-between, and load resistor R_1 matched to the wave resistance of the cable used to feed the optogalvanic signal to an oscilloscope.

The optogalvanic probing of the plasma was effected by means of

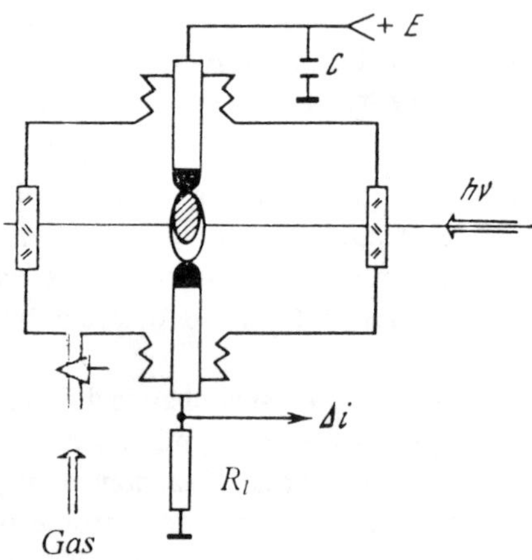

Figure 6.15. Discharge chamber [6.136].

radiation from a tunable dye laser using Rhodamin C dye pumped by the second-harmonic radiation of a YAG laser. The duration of the tunable laser pulse amounted to 10 ns, its power being 40 kW. To improve spatial resolution, radiation was passed into the plasma through a lens 20 cm in focal length. The light-induced plasma conductivity variations were detected by monitoring the discharge current with the oscilloscope or recorded, the pulsed signal being preliminarily integrated.

When the gas pressure was raised in excess of 20 Torr, the glow discharge suffered constriction, the negative glow region and discharge column contracting to 0.2 cm in cross section. The formation of nitrogen dioxide with its characteristic intense coloring was observed in the chamber at that time. Once NO_2 formed, the shape of the optogalvanic signal changed. Typical oscillograph traces of the discharge current response to irradiation of various discharge plasma zones are presented in Figs. 6.16a, b, and c.

The first positive peak was observed both in pure N_2 and O_2 and in their mixtures. Its nature is associated with the photoionization of the atoms from their excited states whose populations in constricted discharge conditions can be significant. The duration of the photoionization pulse is governed by the bulk recombination time and the electron attachment processes.

The second pulse, whose duration was more than two orders of magnitude longer than that of the first, was observed to appear following the plasma-chemical formation of NO_2. The amplitude of this signal grew

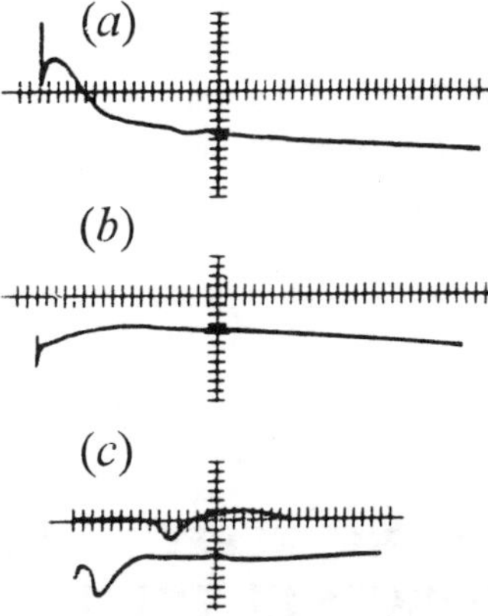

Figure 6.16. (*a*) Optogalvanic signal on the axis of the cathode region of a constricted discharge; (*b*) the signal with the irradiation region displaced 2 mm from the axis,; (*c*) optogalvanic signal in the case of an acoustic perturbation of the discharge plasma. Scan, µs/division: (*a*) and (*b*) – 100;(*c*) – 5.

higher with time to reach saturation in about 50 min. when dynamic equilibrium was established in the gas composition in the enclosed volume of the discharge chamber. With the laser radiation being absorbed in the NO_2 bands, the amplitude and duration of the optogalvanic signal also varied when the laser beam was made to scan the discharge axially and radially. The signal was observed to reach its maximum on the discharge axis in the region of the Faraday dark space (Fig. 6.16*a*), and it diminished as the irradiation region was displaced from the axis toward the periphery of the discharge (Fig. 6.16*b*) and into the positive column region. The formation of the second, longer pulse was apparently due to the processes of convection and diffusion from the laser-excited gas volume containing NO_2.

The action of the laser radiation on the gas surrounding the discharge plasma resulted in the development of a negative discharge current response (Fig. 6.16*c*). But it was superimposed on the positive optogalvanic signal and separated the photoionization and convection-diffusion current responses in time. In Fig. 6.16*c*, the bottom oscillograph trace was recorded when irradiating the discharge in the vicinity of the discharge column, and the top one, with the irradiation region displaced 0.3 cm from its preceding location.

The perturbation propagation velocity far from the discharge amounted to 320 m/s and grew to 500 m/s in the hot gas near the discharge column. This points to an acoustic character of the discharge perturbation due to the radiation heated gas volume. The amplitude of the acoustic signal grew higher with time and reached saturation in 50 min. as in the case of the second positive pulse.

The above examples show that the optogalvanic technique can be used to study the spatial-temporal and optical characteristics of plasma-chemical processes.

6.11. OPTOGALVANIC EFFECT DUE TO PONDERO-MOTIVE FORCES IN THE RADIO-FREQUENCY BAND

The optogalvanic effect considered above is due to changes in the electrical characteristics of the plasma. But the manifestation form of the effect can be associated with the generation of plasma noise caused by the optical induction of ion-sound waves [6.137]. This is due to the action of resonance radiation on the translational motion of the absorbing particles, associated with the spontaneous light pressure force $\mathbf{F}(v)$ given by

$$\mathbf{F}(v) = \frac{\hbar \mathbf{k}_0 \gamma |V|^2}{\left(k_0 v - \Delta\right)^2 + g^2},\tag{6.23}$$

where $\mathbf{k}_0$ is the wave vector, $\Delta = \omega_0 - \omega_{21}$, ω_{21} is the ion quantum transition frequency, ω_0 is the radiation frequency, γ is the excited state decay rate, $g = \sqrt{2|V|^2 + \gamma^2}$ is the optical resonance width with due regard for the field-induced broadening, $V = dE_0/\hbar$ is the Rabi frequency, and E_0 is the electric field amplitude of light. The resonance dependence of the force $\mathbf{F}$ on velocity causes the ions to group in the velocity space in the region $v \sim \Delta/k_0$, where $v = \mathbf{k}_0\mathbf{v}/k_0$ is the ion velocity projection on the radiation propagation direction In that case, a directed structure in the form of a "dip" and a "hump" with a characteristic width of $\delta v \approx g/k_0$ is formed during the characteristic time $T_R = \omega_R^{-1}$ ($\omega_R = (\hbar k_0^2/2m)(2\gamma |V|^2/g^3)$) in the ion velocity distribution function, i.e., there takes place the velocity monochromatization of the ions.

A characteristic feature of the ionic ensemble is associated with the possibility of collective motion. The physical situation of interest to us, where interaction takes place between the internal, translational and collective degrees of freedom, can occur if (a) optical and (b) Cherenkov resonance conditions,

$$(a)\ \Delta \approx \mathbf{k}_0 \mathbf{v} \qquad (b)\ \omega_r(k) \approx \frac{kc_s}{\sqrt{1 + k^2\lambda_e^2}} \approx \mathbf{k}\mathbf{v}\,, \qquad (6.24)$$

can be combined. In the above expressions, λ_e is the electronic Debye radius and $c_s = \sqrt{T_e / m_i}$, ω_r, and $\mathbf{k}$ are the velocity, frequency and wave vector of ion sound waves. Conditions (6.24) can easily be realized by tuning the optical field frequency to the far wing of the Doppler absorption line: $\Delta \leq k_0 c_s$. Because of the conversion of the Landau damping on the ions grouped in the velocity space by the optical field, the combination of conditions (6.24) may give rise to what is knows as the light-induced instability of the ion sound (LIIS).

The scattering of ions by the light-induced fluctuations can radically affect the ion grouping process itself. In a sufficiently rarefied plasma, ion-sound oscillations are slow relative to the radiative relaxation processes, and the change of the translational state and intensity of the LIIS noise is in turn a slow process relative to plasma oscillations, i.e., the conditions for the adiabaticity of interaction between the various degrees of freedom of the plasma are satisfied.

The scattering of ions by plasma fluctuations materially hinders the ion velocity monochromatization process.

In the case of short interaction times, there becomes manifest the above ion velocity monochromatization effect (the appearance of a peak in the ion velocity distribution function). The further intensification of the plasma noise up to a level high above that of thermal noise is accompanied by the smoothing out of the nonequilibrium structure in the distribution function and the formation of a "plateau". In that case, the asymptotic expression for the spectral energy density of the ion-sound noise has the form

$$I(v,t) \approx \frac{1}{\mu}\frac{v^2}{c_s^2}\frac{F(v)}{m_i}\int_{T_R}^t a(t)dt \,. \qquad (6.25)$$

A remarkable fact is the reproduction of the radiative profile of the force $F(v)$ by the spectrum of the plasma noise $I(v, t)$.

It has been well known that the optical absorption spectrum in a weak light field is governed by the imaginary part of the linear optical susceptibility $\chi(\omega)$. In the case under consideration, one may write down the following expression for a strong spontaneous light pressure:

$$\mathbf{F}(v) = (p\nabla\widehat{E}^* + \mathrm{c.c.}) = 2\mathbf{k}_0\mathrm{Im}\chi(\omega_0 - k_0 v)\left|E_0^2\right|.$$

Thence

$$I(v) \propto \mathrm{Im}\chi(\omega_0 - k_0 v),$$

i.e., the plasma noise spectrum reproduces the optical absorption spectrum.

In that case, the possible multiple-resonance structure $\mathrm{Im}(\chi)$ associated with the presence of close quantum transitions with overlapping Doppler profiles is allowed in the spectrum of the above-thermal plasma noise at a fixed (!) optical frequency.

This effect can form the basis for a new method for the sub-Doppler optogalvanic spectroscopy of ions by way of plasma noise detection.

What is important is that the signal (plasma noise at a plasma concentration of $n < 10^9$–10^{10} cm^{-3} of interest to us) is fixed in the radio-frequency range convenient for detection.

Note that collective phenomena fairly similar in nature to those under discussion have already been observed in the first experiment on the laser cooling of a relativistic beam of $^7\mathrm{Li}^+$ ions [6.138]. These phenomena have manifested themselves in the form of a narrow peak in the spectrum of noise induced by the ion beam on the electrodes surrounding it.

REFERENCES

6.1.   P Camus, *J. de Phys.* **4**, C7: 87–106 (1983).

6.2.   N. Barberi, N. Beverini, A. Sasso, *Rev. Mod. Phys.* **62**, No. 3: 603–645 (1990)

6.3.   P. Camus, M. Dieulin, A. E. Himdly, M. Aymar, *Phys. Scripta* **27**: 125–136 (1983).

6.4.   C. J. Lorenzen, K. Niemax, L. R. Pendrill, *Opt. Commun.* **39**: 271 (1980).

6.5.   K. Niemax, L. R. Pendrill, *J. Phys. B* **13**: 461–465 (1980).

6.6.   L. R. Pendrill, K. Niemax, *J. Phys. B* **15**: 147–151 (1982).

6.7.   C. J. Lorenzen, K. Niemax, *Phys. Scripta.* **27**: 300–305 (1983).

6.8. C. J. Lorenzen, K. Niemax, *J. Phys. B* **15**: 139–145 (1982).

6.9. K. Niemax, *Appl. Phys.* **32**: 59–62 (1983).

6.10. C. J. Lorenzen, K. Niemax, *J. Phys. A* **311**: 249–250 (1983).

6.11. C. J. Lorenzen, K. Niemax, *Opt. Commun.* **43**: 26–30 (1982).

6.12. L. Ph. Roesch, *Opt. Commun.* **44**: 259–261 (1983).

6.13. R. A. Keller, R. Engleman, E. F. Zalewski, *JOSA* **69**: 738 – 742 (1979).

6.14. P. Pianarosa, Y. Demers, J. M. Gague, *J. Opt. Soc. Amer.* **1**: 704–709 (1984).

6.15. B. Barberi, N. Beverini, M. Galli *et al.*, *Nuovo Chim. D* **4**: 172–180 (1984).

6.16. H. Rinneberg, J. Neukammer, *J. de Phys.* **44** C7: 177–191 (1983).

6.17. J. E. Lawler, A. I. Ferguson, J. E. M. Goldsmyth *et al.*, *Phys. Rev. Lett.* **42**: 1046–1049 (1979).

6.18. R. Brigand, A. Timmerman, *J de Phys.* **44**, C7: 137–148 (1983).

6.19. P. Penent, C. Chardonnet, D. Delande *et al.*, *J de Phys.* **44**, C7: 193–208 (1983).

6.20. J. P. Lemoigne, J. P. Grandin, X. Husson, H. R. F. Kucal, *J de Phys.* **44**, C7: 209–216 (1983).

6.21. T. Caesar, J. L. Henlly, *Opt. Comm.* **45**: 258–260 (1983).

6.22. A. Wada, Y. Adachi, C. J. Hirose, *Chem. Phys.* **86**: 5904 – 5908 (1987).

6.23. A. Wada, Y. Adachi, C. J. Hirose, *J. Chem. Phys.* **90**: 6645–6650 (1987).

6.24. R. C. Webster, C. T. Rettner, *Laser Focus* **19**: 41–52 (1983).

6.25. C. B. Collins, *J de Phys.* **44**, C7: 395–409 (1983).

6.26. R. Vasudev, R. N. Zare, *J. Chem. Phys.* **76**: 5267–5270 (1982).

6.27. K. Miyazaki, H. Scheingraber, C. R. Vidal, *Pys. Rev. A* **28**: 2229–2244 (1983).

6.28. T. Suzuki, M. Kakimoto, *J. Mol. Spectr.* **93**: 423–432 (1982).

6.29. C. Hameau, J. Wascat, D. Dangoisse, P. Glorieux, *Opt. Commun.* **49**: 423–428 (1984).

6.30. C. R. Vidal, *Opt. Lett.* **5**: 158–159 (1980).

6.31. D. Popescu, M. L. Pascu, C. B. Collins *et al.*, *Phys. Rev A* **8**: 1666–1672 (1973).

6.32. C. B. Collins, S. M. Curry, B. W. Johnson *et al.*, *Phys. Rev. A* **14**: 1662–1671 (1976)

6.33. M. Y. Mirza, W. W. Duley, *Proc. R. Soc. London. A* **364**: 255–263 (1978).

6.34. M. Y. Mirza, W. W. Duley, *Opt. Commun.* **28**: 179–182 (1979).

6.35. M. A. Chelleohmalzadeh, C. B. Collins, *Phys. Rev. A* **19**: 2270–2276 (1979).

6.36. C. B. Collins, F. W. Lee, H. Golbani *et al.*, *J. Chem. Phys.* **75**: 4852–4863 (1981).

6.37. P. K. Schenk, W. G. Mallard, J. C. Travis, K. S. Smith, *J. Chem. Phys.* **69**:5147–5150 (1978).

6.38. D. Feldman, *Opt. Commun.* **29**: 67–72 (1979).

6.39. T. Suzuki, *Opt. Commun.* **38**: 364–368 (1981).

3.40. C. T. Rettner, C. R. Webster, R. N. Zare, *J. Phys. Chem.* **85**: 1105–1107 (1981).

6.41. C. Demuynck, J. L. Destombles, *IEEE J. Quant. Electr.* **17**: 575–577 (1981).

6.42. I. M. Beterov, N. V. Fateyev, *Opt. Commun.* **40**: 425–429 (1982).

6.43. I. M. Beterov, N. V. Fateyev, in: Lazernye sistemy (Laser systems) (Novosibirsk, 1982): 121–142) (in Russian).

6.44. J. Pfaff, M. Begemann, R. J. Saykally, *Mol. Phys.* **52**: 541 – 566 (1984).

6.45. Wang Yu-min, Gui Zhen-xing, Zhang Shun-yi, *Acta Optica Sinica* **3**: 797–804 (1983).

6.46. N. K. Zaitsev, N. Ya. Shaparev, *ZhTF* **55**: 218–220 (1985).

6.47. R. C. Webster, R. T. Mensies, *J. Chem. Phys.* **78**: 2121–2128 (1983).

6.48. R. E. Munchausen, R. D. May, G. W. Hill, *Opt. Commun.* **48**: 317–321 (1984).

6.49. S. W. Kim, W. E. Jones, *Spectr. Lett.* **18**: 767–779 (1985).

6.50. S. Moffat, A. L. S. Smith, *Opt. Commun.* **37**: 119 (1981).

6.51. M. C. Scolnik, *IEEE J. Quant. Electr.* **6**: 139 (1970).

6.52. J. Jirmann, W. Durr, *Electron. Lett.* **18**: 69–71 (1982).

6.53. A. L. S. Smith, S. Moffat, *Opt. Commun.* **30**: 213–218 (1979).

6.54. N. J. Kavaya, R. T. Mensies, U. P. Oppenheim, *IEEE J. Quant. Electr.* **18**: 19–21 (1982).

6.55. R. Walkup, R. W. Dreufus, Ph. Avoris, *Phys. Rev. Lett.* **50**: 1846–1849 (1983).

6.56. K. C. Smyth, W. G. Mallard, *J. Chem. Phys.* **77**: 1779–1787 (1982).

6.57. W. B. Bell, A L. Bloom, *J. Appl. Phys.* **32**: 906–909 (1961).

6.58. A. C. Tam, *IEEE Trans. Plasma Sci.* **10**: 252–256 (1982).

6.59. O. I. Matveev, N. B. Zorov, Yu. Yu. Kuzyakov, *Talanta* **27**: 907–908 (1980).

6.60. S. W. Downey, N. S. Nogar, *Analyt. Chem.* **57**: 13–16 (1985).

6.61. S. N. Raikov, *Razvitiye analiticheskikh vozmozhnostei vnutrirezonansnoi spektroskopii s primeneniyem lazerov na krasitelyakh impulsnogo deistviya* (Development of analytical capabilities of the intracavity spectroscopy technique using pulsed dye lasers) (Minsk: In-t Fiziki AN SSSR, 1986) (in Russian).

6.62. E. F. Zalevski, R. A. Keller, C. T. Apel, *Appl. Opt.* **20**: 1584–1590 (1981).

6.63. D. Apostol, C. Blanary, *Rev. Roum. Phys.* **27**: 581–585 (1982).

6.64. A. Sasaki, S. Usimaru, T. Hayashi, *Jap. J. Appl. Phys.* **24**: 593–600 (1984).

6.65. *Laser Focus* **19**: 147 (1983).

6.66. M. Schneider, A. Hinz, A. Groh, K. Evenson, W. Urban, *Appl. Phys.* **B44**: 241 (1987).

6.67. G.D. Lobov, V.V. Shtykov, V. I. Bogatkin, L. V. Drugov, *Radiotekhnika i Elektronika* **6**: 1246–1251 (1972).

6.68. R. B. Green, R. A. Keller, G. G. Luther *et al.*, *IEEE J. Quant. Electr.* **13**: 63–64 (1977).

6.69. V. S. Burakov, A. A. Gvozdev, P. Ya. Misakov *et al.*, in: *Tez. dokl. XIX Vsesoyuz. siezda po spektroskopii* (Proc. XIX All-Union Congress on spectroscopy) (Tomsk, 1983), vol. 1: 271. (in Russian).

6.70. C. A. Hamer, J. P. Sponhower, *Appl. Spectr.* **38**: 212–214 (1984).

6.71. J.-T. Shy, T. C. Yen, *Opt. Commun.* **60**: 306–308 (1986).

6.72. D. C. King, P. K. Schenk, K. C. Smyth, J. C. Travis, *Appl. Opt.* **16**: 2617–2619 (1977).

6.73. J. Jin, C. Jin, S. Wang, *Chinese Phys.* **3**: 430–432 (1983).

6.74. R. Avarmaa, *Izv. AN ESSR .Fizika. Matematika* **33**: 333–338 (1984).

6.75. N. J. Dovichi, D. S. Moore, R. A. Keller, *Appl. Opt.* **21**: 1468–1472 (1982).

6.76. J. R. Nestor, *J de Phys.* **44**, C7: 387–388 (1983).

6.77. C. R. Webster, *Appl. Opt.* **21**: 2298–2300 (1982).

6.78. A. Ben-Amar, R. Shuker, G. Erez, *Appl. Phys. Lett.* **38**: 763–765 (1981).

6.79. I. M. Beterov, N. V. Fateev, *Opt. Commun.* **40**: 425–429 (1982).

6.80. I. M. Beterov, N. V. Fateev, *Optika i Spektroskopiya* **54**: 978–986 (1983).

6.81. S. Avriller, J. P. Schermann, *Opt. Commun.* **19**: 87–91 (1976).

6.82. I. M. Beterov, N. V. Fateev, in: *Ispol'zovaniye obiyomnoi i poverkhnostnoi otritsatel'noi ionizatsii dlya detektirovaniya molekul, vozbuzhdayemykh lazernym izlucheniyem* (The use of bulk and surface negative ionization to detect molecules excited by laser radiation) (Novosibirsk, 1982): 121–142 (in Russian).

6.83. I. M. Beterov, V. A. Kurochkin, N. V. Fateev, *Zhurn. Khim. Fiz.* **1**: 357–366 (1982).

6.84. I. M. Beterov, N. V. Fateev, in: B. M. Smirnov, ed., S*tolknovitelnaya ionizatsiya elektronno-vozbuzhdennykh atomov* (Collision ionization of electronically excited atoms) (Moscow: Energoatomizdat, 1987): 40–73 (in Russian).

6.85. J. C. While, R. R. Freeman, P. F. Liao, *Opt. Lett.* **5**: 120–122 (1980).

6.86. R. Hotop, K. Niemax, J. Richter, K. H. Weber, *Opt. Commun.* **36**: 35–38 (1982).

6.87. M. Tokashima, J. Fujita, O. Yamazali *et al.*, *Proc. XI Conf. on Gas Discharges and their Applications* (Tokio, 1995).

6.88. K. H. Weber, K. Niemax, *J. Phys A* **307**: 13–18 (1982).

6.89. K. Niemax, *Appl. Phys. B* **32**: 59–62 (1983).

6.90. R. A. Zhitnikov, in: *Tez. obzornykh dokl. VI Vsesoyuz. konf. po fizike nizkotemperaturnoi plazmy* (Proc. VI All-Union conference on low-temperature plasma physics) (Leningrad, 1983): 83 (in Russian).

6.91. F. O. Shimizu, K. Sasaki, K. Ueda, *Jap. J. Appl. Phys.* **22**: 1144–1151 (1983).

6.92. A. V. Kirillin, K. A. Khodakov, *TVT* **22**: 1223–1225 (1983).

6.93. I. T. Yakubov, K. A. Khodakov, *TVT* **22**: 14–19 (1984).

6.94. Y. Yasuda, N.Sokabe, A.Murai, *Opt. Commun.* **55**: 319–325 (1985).

6.95. L. Yin, Q. Hu, H. Su, F. Lin, *Chinrse Phys.* **4**: 691–695 (1984).

6.96. R. Klein, P.McGinnis, R.Leone, *Chem. Phys. Lett.* **100**: 475–478 (1983).

6.97. H.A. Bachor, P.J. Manson, R.J. Sandeman, *Opt. Commun.* **43**: 337–342 (1982).

6.98. D.J. Jackson, E. Arimondo, J.E. Lawler, T.W. Hänsch, *Opt. Comm.* **33**: 51–55 (1980).

6.99. T. F. Johnson, *Laser Focus* **14**: 58 (1978).

6.100. E. M. van Veldhuisen, *The hollow Cathode Glow Discharge Analyzed by optogalvanic and Other Studies: Thesis* (Eindhoven, 1983).

6.101. H. O. Rehrens, G. H. Guthöhrlein, *J. de Phys.* **44**, C7: 149 – 168 (1983).

6.102. V. K. Otorbayev, V. N. Ochkin, N. G. Preobrazhensky *et al.*, *ZhETF* **81**: 1626–1638 (1981).

6.103. C. R. Webster, R. T. Mensies, *Proc. SPIE* **413**: 152–157 (1984).

6.104. K. E. Grundberg, G. A. Hebuer, J. T. Verdyen, *Appl. Phys. Lett.* **44**: 299–302 (1984).

6.105. R. A. Gottscho, C. E. Gaebe, *IEEE Trans. Plasma Sci.* **PS-14**: 92–102 (1986).

6.106. J. Kramer, *J. Appl. Phys.* **60**: 3072–3080 (1986).

6.107. C. M. Ferreira, J. Loureiro, M. Pinheiro, P. A. Sa, in: *Nonequilibrium Processes in Partially ionized Gases*, Ed. by M. Capitelly and J. H. Bardsley. NATO ASI Series (New York: Plenum Press, 1989).

6.108. T. Suzuki, T. Kasuya, *Appl. Phys.* **B51**: 374–378 (1990).

6.109. D. K. Doughty, S. Salin, J. E. Lawler, *Phys. Lett.* **103** 41–44 (1984).

6.110. D. K. Doughty, J. E. Lawler, *Appl. Phys. Lett.* **45**(6): 611 – 613 (1984).

6.111. B. N. Ganguly, A. Garscadden, *Appl. Phys. Lett.* **46**: 540 – 542 (1985).

6.112. B. N. Ganguly, A. Garscadden, *Phys. Rev.* **A32**: 2544–2545 (1985).

6.113. H. Debontride, J. Deroruard, P. Edel *et al.*, *Phys. Rev.* **A40**: 5208–5219 (1989).

6.114. J. E. Lawler, D.A. Doughty, in: *Advances in Atomic, Molecular and Optical Physicsl* **34** (Academic Press, 1994): 171–206.

6.115. D. Kumar, R. R. Zinn, T. D. Armstrong, S. R. McGlynn, *J. Phys. Chem.* **99**, No. 19: 7530–7536 (1995).

6.116. R. B. Green, R. A. Keller, P. K. *et al.*, *J. Am. Chem. Soc.* **98**: 1517–1518 (1976).

6.117. R. A. Keller, E. F. Zalewski, *Chinese Phys.* **5**: 973–979 (1985).

6.118. P. K. Schenk, K. C. Smyth, *J. Opt. Soc. Am.* **68**: 626–627 (1978).

6.119. M. N. Reddy, Rao-Gin, *Phys. C* **C150**: 457–465 (1988).

6.120. C.A. Van Dijk, P.J.Th. Zeegers, G. Hienhues, C.Th. J. Alcemade, *J. Quant. Spectr. Rad. Transf.* **20**: 55–63 (1978).

6.121. Yu. Yu. Kuzyakov, N. B. Zorov, V. I. Chaplygin *et al.*, *J. Phys.* **44**: 335–343 (1983).

6.122. G. C. Turk, J. C. Travis, J. R. De Voe, *J. de Phys.* **44**, C7: 301–309 (1983).

6.123. J. C. Travis, G. C. Turk, R. E. Green, *Anal. Chem.* **54**: 1006–1018 (1982).

6.124. Th. Berthoud, P. Camus, N. Drin, J. L. Stehle, *J. de Phys.* **44**, C7: 389–391 (1983).

6.125. V. N. Chaplygin, *Lazernyi atomno-ionizatsionnyi metod opredeleniya sledov kaliya, rubidiya i tseziya pri atomizatsii proby v plameni: Thesis* (Laser atomic-ionization technique for determining potassium, rubidium, and cesium traces in a flame) (Moscow: MGU, 1984) (in Russian).

6.126. D. E. Murnick, M. J. Colgan, H. P. Lie, D. Stoneback, *SPIE Proc.* **1768**: 454–463 (1996).

6.127. S. Moffat, A. L. S. Smith, *J. de Phys.* **17**: 59–70 (1984).

6.128. E. Arimondo, M. G. De Vito, K. Ernst, M. Inguscio, *Opt. Commun.* **9**: 530–532 (1984).

6.129. E. Arimondo, M. G. De Vito, K. Ernst, M. Inguscio, *J. de Phys.* **44**: 267–270.

6.130. N. Ya. Shaparev, N. K. Zaitsev, V. A. Pushkaryov, *Dinamichesky optogal'vanichesky effekt v plazme neona* (Dynamic optogalvanic effect in a neon plasma) (Krasnoyarsk: Preprint In-ta Fiziki No. 274Ф, 1984) (in Russian).

6.131. W. B. Bridges, *JOSA* **68**: 352–360 (1978).

6.132. M. J. Kushner, *Appl. Opt.* **22**: 1970–1975 (1983).

6.133. N. Ya. Shaparev, *ZhTF* **49**: 2229–2231 (1979).

6.134. A. C. Predvoditelev, E. V. Rozhdestvensky, E. V. Stupochenko *et al.*, *Tablitsy termodinamicheskikh funktsy vozdukha* (Tables of thermodynamic functions of air) (Moscow: VTs AN SSSR, 1962).

6.135. V. D. Rusanov, A. L. Fridman, G. V. Sholin, *UFN* **134**: 185–203 (1981).

6.136. N. K. Zaitsev, N. Ya. Shaparev, G. V. Sholin, *ZhTF* **55**: 218–220 (1985).

6.137. I. V. Krasnov, N. Ya. Shaparev, *Optika i Spektroskopiya* **80**: 577–580 (1996).

6.138. S. Schroder, R. Klein, N. Boss *et al.*, *Phys. Rev. Lett.* **64**: 2901–2904 (1990).

Chapter 7
Conclusion

The optogalvanic effect with its fairly long history has been going through its second birth ever since the very first works appeared on the use of lasers for its observation. The vigorous growth of interest in this subject and increasing list of publications have made the optogalvanic technique a new fundamental method for studying plasma media. The effect is also being used to actively influence various objects, for example, to stabilize the frequency and power output of lasers. There are other interesting possible applications that are outside the scope of this book, such as its use in communication systems and high-power electric circuits [7.1]. The optogalvanic effect has become the object of discussion at specialized conferences. As far as its range of aspects and applications is concerned, the optogalvanic effect in plasmas and gases may be compared with the photoelectric spectroscopy of impurities in semiconductors [7.2] and other well-known applications of the photoconductivity of solids [7.3].

An important merit of the optogalvanic detection technique is its high sensitivity and selectivity in the quantum states of particles. It can therefore be used, in conjunction with the traditional methods, to study elementary processes and to diagnose atomic and molecular structures, too. The potential capabilities of the technique are tremendous.

At the same time, the question as to the sensitivity and ultimate signal/noise ratio of the optogalvanic technique in comparison with those of the direct absorption measurement, laser-induced fluorescence, optoacoustic measurement, and other techniques needs to be analyzed in greater detail. In this respect, some general considerations have been put forward in [7.4], the advantages of the optogalvanic technique being emphasized by the authors. The analytical noise-associated limitations of the technique have obviously not been adequately studied, except for some particular cases of hollow-cathode discharge [7.5], flame [7.6], and some positive column conditions [7.7, 7.8].

In many situations the optogalvanic techniques described in this book come close to the other, experimentally different research methods that are now being intensely developed. These include in particular multistep photoionization spectroscopy techniques [7.9] whose unique capabilities enable one to detect single atoms and molecules. The methods with the direct detection of ions [7.10, 7.11] that have already been mentioned should apparently be placed into the same category.

Recent attention has been focused on entirely new and at first glance unexpected manifestations and applications of the optogalvanic effect. Specifically, a scheme has been suggested in [7.12] for isolating plasma particles from their confinement walls by creating an optical potential barrier under biharmonic irradiation at frequencies close to the resonance frequencies of the ions. Isolation in that case proves to be selective as to the ionic species. Further development of the optical isolation idea has pointed to the possibility that the plasma can be cooled deeply, down to the state of crystallization [7.13–7.15].

The optogalvanic effect considered in this monograph is being studied under laboratory conditions and is associated with the action of laser radiation on the plasma. We would like to draw the reader's attention to the global optogalvanic effects existing in nature. Worthy of notice among them are the so-called sudden ionospheric disturbances well-known to the specialists in the physics of ionosphere [7.16]. Their origin is as follows. The solar radiation reaches the Earth's atmosphere and ionizes the particles contained in its upper part to form conductive layers – the ionosphere – therein. The solar wind flowing over the Earth's magnetosphere produces an electric potential difference between its morning and evening sides, which is transmitted over the geomagnetic lines of force into the ionosphere to generate electric currents therein, which flow in on the morning side and out on the evening side. This in turn changes the Earth's magnetic field. When a flare occurs on the Sun, the flow of ionizing radiation (in the UV and X-ray regions) is intensified, so that within a few minutes the concentration of electrons in the D and E regions of the ionosphere increases tens of times. The increase of the conductance of the ionosphere leads to an increase in the current, which manifests itself in a stepwise change in the geomagnetic field.

The nature of the optogalvanic effect, as simple as it is, combines with the complex, intricately branched mechanisms underlying the formation of the specific features of the optogalvanic signal, especially under nonequilibrium conditions. So far, a quantitative description has only been possible for some simple cases and objects of study. The phenomenon calls for further experimental and theoretical model investigations, which will undoubtedly open up new possibilities and facilitate the realization of those which have already taken shape.

ACKNOWLEDGMENTS

The authors of the book are deeply thankful to the Editors of Journals and authors of publications cited for their kind permission to use some graphic materials published therein, namely:

Optics Communications **39** (1981), Fig. 3, Ref. No. [2.7]; **43** (1982), Figs. 3 through 5, Ref. No. [6.97]; **44** (1983), Fig. 3, Ref. No. [1.21];

Physics Letters A **103** (1984), Figs. 2 and 3, Ref. No. [6.109];

Journal of Physics D **17** (1984), Figs. 3 and 8, Ref. No. [2.34];

Applied Physics Letters **44** (1984), Fig. 3, Ref. No. [6.104];

Journal of Optical Society of America **1** (1984), Fig. 2, Ref. No. [6.14];

Applied Optics **22** (1983), Figs. 1 through 3, Ref. No. [6.132];

Physical Review A **22** (1980), Fig. 2, Ref. No. [2.5]; **28** (1983), Figs. 2 and 5*a*, Ref. No. [2.13]; **36** (1987), Fig. 2, Ref. No. [3.55];

Journal of Chemical Physics **78** (1983), **78**, Fig. 3*b*, Ref No. [1.19];

IEEE Journal of Quantum Electronics **14** (1978), Figs 1*a* and 3, Ref. No. [2.6];

· *Analytical Chemistry* **56** (1984), Ref. No. [4.16].

REFERENCES

7.1. J. E. Lawler, D. K. Doughty, *Digest of Technical Papers, 4th IEEEE International Pulsed Power Conference* (1983): 33–36.

7.2. Sh. M. Kogan, T. M. Lifshits, *Phys. Stat. Solidi.* **39a**: 11–39 (1977).

7.3. R. Bube, *Potoconductivity of Solids* (New York, 1960).

7.4. R. C. Webster, C. T. Rettner, *Laser Focus* **18**: 41–52 (1983).

7.5. R. A. Keller, E. F. Zalewski, *Appl. Opt.* **19**: 3301–3305 (1980).

7.6. O. I. Matveyev, *Optika i Spektroskopiya* **63**, No. 5: 1171 – 1174 (1987).

7.7. N. S. Kopeika, J. Rosenbaum, *IEEE Trans. Plasma. Sci.* **PS4**: 51–61 (1976).

7.8. N. S. Kopeika, *IEEE Trans. Plasma. Sci.* **PS6**: 261–270 (1978).

7.9. V.S. Letokhov, *Lazernaya fotoionizatsionnaya spektroskopiya* (Laser photoionization spectroscopy) (Moscow: Nauka, 1987) (in Russian).

7.10. I. M. Beterov, N. V. Fateev, *Opt. Commun.* **40**: 425–429 (1982).

7.11. I. M. Beterov, N. V. Fateev, *Optika i Spectroskopiya* **54**: 978–986 (1983).

7.12. A. P. Gavrilyuk, I. V. Krasnov, Trapeznikov, N. Ya. Shaparev, *Izv. VUZov. Fizika*, No. 5: 78–81 (1998).

7.13. A. P. Gavrilyuk, I. V. Krasnov, N. Ya. Shaparev, *Pis'ma ZhETF* **63**, No. 5: 316–321 (1996).

7.14. A. P. Gavrilyuk, I. V. Krasnov, N. Ya. Shaparev, *Pis'ma ZhETF* **23**, No. 2: 28–32 (1997).

7.15. A. P. Gavrilyuk, I. V. Krasnov, N. Ya. Shaparev, in: *Digest. The Second International Symposium on Modern Problems of Laser Physics* (Novosibirsk, Russia, 1997): 53–54.

7.16. B. E. Bryunelli, A. A. Namchaladze, *Fizika ionosfery* (Physics of ionosphere) (Moscow: Nauka, 1988) (in Russian).

Index